Lab Manual
to accompany

The Commercial Greenhouse

Third Edition

Lab Manual
to accompany
The Commercial Greenhouse
Third Edition

James W. Boodley
Steven E. Newman

Contributing Authors
Elmer L. Cooper
Charles Griner
Catherine A. Teare Ketter
Carroll L. Shry, Jr.
Elizabeth Strum
Walter York

DELMAR
CENGAGE Learning™

Australia • Brazil • Japan • Korea • Mexico • Singapore • Spain • United Kingdom • United States

DELMAR
CENGAGE Learning™

Lab Manual to Accompany
The Commercial Greenhouse, 3E
James W. Boodley and,
Steven E. Newman

Vice President, Career and Professional Editorial: Dave Garza

Director of Learning Solutions: Matthew Kane

Senior Acquisitions Editor: David Rosenbaum

Managing Editor: Marah Bellegarde

Product Manager: Christina Gifford

Editorial Assistant: Scott Royael

Vice President, Career and Professional Marketing: Jennifer McAvey

Marketing Director: Deborah Yarnell

Marketing Coordinator: Jonathan Sheehan

Production Director: Carolyn Miller

Production Manager: Andrew Crouth

Content Project Manager: Katie Wachtl

Senior Art Director: Dave Arsenault

Technology Project Manager: Chris Catalina

Production Technology Analyst: Thomas Stover

Library of Congress Control Number: 2008925689

ISBN-13: 978-1-4180-3083-4

ISBN-10: 1-4180-3083-X

Delmar
Executive Woods
5 Maxwell Drive
Clifton Park, NY 12065
USA

Cengage Learning is a leading provider of customized learning solutions with office locations around the globe, including Singapore, the United Kingdom, Australia, Mexico, Brazil, and Japan. Locate your local office at **www.cengage.com/global**

Cengage Learning products are represented in Canada by Nelson Education, Ltd.

To learn more about Delmar, visit **www.cengage.com/delmar**

Purchase any of our products at your local college store or at our preferred online store **www.cengagebrain.com**

Printed in the United States of America
2 3 4 5 6 24 23 22 21 20

CONTENTS

SECTION

1

The Industry—Scope and Development

Notes

Name: ______________________ Date: ______________

Exercise 1-1: Personal Data Sheet or Resume

(by Carroll L. Shry, Jr.)

Purpose

To prepare a personal data sheet or resume.

Materials

Pen or pencil

Procedure

Using the sample resume as an example, prepare your own resume by filling in the spaces in the blank resume that follows it.

RESUME

Steve C. Cover

468 Hopemont Street
Woodsboro, Maryland 21798
301-663-8007

Career Objective:	Seek a position with a commercial landscaping company	
Education:	Senior at Delmar High School, to graduate in May, 2009. Majoring in Landscape and Nursery Management	
Subjects Studied:	Algebra:	1 semester
	AgriScience:	2 semesters
	Biology:	2 semesters
	Chemistry:	1 semester
	Consumer Economics:	1 semester
	Cooperative Work Experience:	2 semesters
	English:	4 semesters
	Landscape and Nursery Production:	2 semesters
Student Activities:	President of the FFA Chapter Treasurer of the Senior Class Members of the School Band	
Work Experience:	September 2007 to present Blue Mountain Landscape Company 133 Dogwood Lane, Frederick, Maryland 21701	
References:	1. Mr. Ron Smith 7999 River Road, Walkersville, Maryland 21793 2. Mrs. Linda Summers 243 Nolands Ferry Road, Tuscarora, Maryland 21701	

Name: ______________________________ Date: ______________

Resume

Name: ______________________________

Address: ______________________________

Telephone: ______________________________

Career Objective: ______________________________

Education: ______________________________

____

Subjects Studied: ______________________________

____

____

____

____

____

____

Student Activities: ______________________________

____

____

____

____

____

Work Experience: ______________________________

____

____

References: ______________________________

____

____

Name: ______________________________ Date: ______________

Exercise 1-2: Job Application

(by Carroll L. Shry, Jr.)

Purpose

To complete a job application.

Materials

Pen or pencil

Procedure

Complete the following job application.

Name: ______________________________ Date: ______________

Application for Employment

All applicants will be considered for employment without regard to race, religion, color, sex, national origin, age, marital status, disability, sexual orientation, veteran status or any other status protected by law. We are an Equal Oppoutunity Employer.

Personal Information *(please print)*

Date: ______________

Name ______________________________ Soc. Sec. # ______________

Address ______________________ City __________ State ______ Zip ______

Telephone No. ______________ Referred by: ______________________

Are you legally eligible for permanent employment in the U.S.? ______________

Position(s) applied for ______________________ Full Time ❑ Part Time ❑

Seasonal Help only ❑ Yes ❑ No

If part time, specify days/hours available ______________________

Rates of pay desired $ ______ per ______

Have you worked for us before? ______ If YES, when? __________ Position __________

Indicate special qualifications or skills ______________________

Education

Name and Location of School	Course of Study	Did You Graduate?	If You Didn't Graduate Years Completed or Attended
Elementary			
High School			
College	Major Degree		
Other			

Special Questions

Are you over 18 years of age? ______ If NO, state your age: _____ (employment subject to legal age verification)

** It is unlawful in MA & MD to require or administer a lie detector test as a condition of employment or contined employment. An employer who violages this law shall be subject to criminal penalties and civil liability.

(continued)

Name: ______________________________ Date: ______________

Prior Employment *(start with most recent employer; you may also include volunteer work)*

Employer:	Phone:	From:	To:
Address:	City, State, Zip:	Position:	
Duties:		Supervisor's Name:	
		Starting Salary/Wages:	
Reason for leaving:		Final Salary/Wages:	
Employer:	Phone:	From:	To:
Address:	City, State, Zip:	Position:	
Duties:		Supervisor's Name:	
		Starting Salary/Wages:	
Reason for leaving:		Final Salary/Wages:	
Employer:	Phone:	From:	To:
Address:	City, State, Zip:	Position:	
Duties:		Supervisor's Name:	
		Starting Salary/Wages:	
Reason for leaving:		Final Salary/Wages:	

Military Service

Branch of Service	From	To	Rank and Duties	Date Discharged

Business References

(Please supply us with three business references, preferably current or former supervisors. If unable to supply three supervisors please give coworkers or subordinates.)

Name	Address	Years Known	Telephone

I authorize investigation of all statements contained in this application. I understand that misrepresentation, falsification or omission of facts called for is cause for dismissal. Further, I understand and agree that my employment is for no definite period and may be terminated at any time without any previous notice.

Signature: ______________________________

REMARKS:

SECTION

2

Types of Plant Growing Structures

Notes

Name: ______________________________ Date: ______________

Exercise 2-1: Greenhouse Pipe Frame

(by Elmer L. Cooper)

Purpose

To construct a greenhouse pipe frame.

Materials

	Quantity	Dimensions	Description
Wood	4	1″ × 4″ × 12′	Ground board
Wood	4	2″ × 4″ × 8′	Door frame
Plywood, exterior grade	2	4′ × 8′ A/C	Doors
PVC pipe Sch. 40	7	1″ × 20′	Bows and braces
PVC pipe Sch. 40	8	1½″ × 18″	Standards
#9 wire	10′	#9 × 10′	Straps
Hex bolts and nuts	12	¼″ × 2″	Attach bows
Plastic film	1	20′ × 30′	6 mil thick

Miscellaneous: door hinges, staples, nails, and paint

Procedure

Construction Procedure for Greenhouse Pipe Frame

1. Prepare a level site and ensure adequate drainage.
2. Square the structure. Drive or set in concrete the 18″ × 1½″ PVC standards so only 4″ are visible.
3. Drive or set remaining standards on 4′ centers down each side until only 4″ are visible. The tops of all standards should be level.
4. Mark four 1″ PVC pipes with a line 6″ from each end and at the midpoint of each pipe.
5. Place the 1″ × 4″ ground boards around the outside of all the standards.
6. Insert one end of a 1″ PVC pipe into the first corner standard to the 6″ mark and drill through the ground board, bow, and standard. Insert a ¼″ bolt and tighten.
7. Insert the opposite end of the 1″ PVC pipe into the corner standard across from the first one and secure as described in #6.
8. Complete installation of all bows.
9. Cut 1″ PVC pipe 12′ long and mark every 4′. Hold this purlin up to the inside of the arched bows. Drill and insert ¼″ bolts with purlin marks aligned with the center of all bows. Saw off any protruding bolts to prevent head injuries.
10. Take 1″ PVC pipe and bolt end about 4′ high on the inside of end bow. Allow the pipe to triangulate down to the ground board. Bolt the pipe to all bows that it passes and to the ground board. This is a wind brace and is an important member of the greenhouse. Be sure bows are plumb.

Name: ______________________________ Date: ______________

11. Construct the end walls and door frames as illustrated.

12. Drape plastic across the framework and staple the leading edge to the door frame. Proceed down each side, pulling the plastic tight to the ground board. Roll the edge of the plastic several times and staple the roll to the wooden members. Finish with the trailing end attached to the opposite door frame.

13. Doors, hinges, and utilities are constructed according to standard practices.

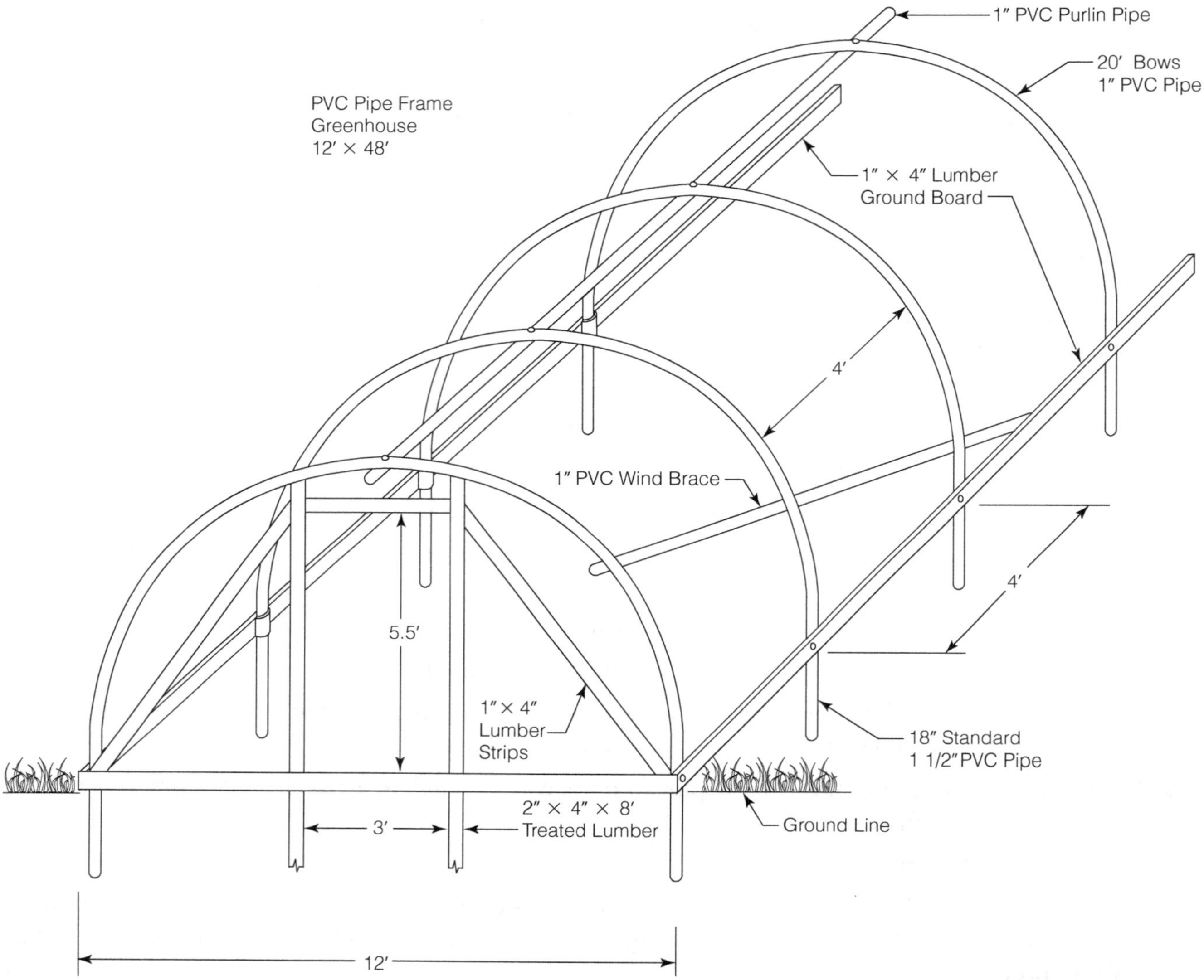

Courtesy University of Maryland at Princess Anne, Dr. Thomas Handwerker

Name: ______________________________ Date: ______________

Exercise 2-2: Hydroponics Model—Tube Culture

(by Elmer L. Cooper)

Purpose

To construct a hydroponics model tube culture.

Materials

	Quantity	Dimensions	Description
Wood	12′	1″ × 4″	Frame
Wood	2′	1″ × 8″	Upper support
Wood	2′	1″ × 6″	Lower support
PVC pipe Sch. 40	20′	2″	Growing tube
PVC cap	5, 1	2″, 1″	Caps
Plastic pipe	2′	1″	Manifold
Plastic tubing and fittings	10′	5/16″	Supply line
Header tubing	4′	1/4″	Header tubing
Submersible pump	1	Aquarium size	Pump
Reducing coupling	1	1″ <> 5/16″	Connection
Plastic pail	1	5 gallon	Stock tank

Miscellaneous: wood screws, paint and nails

Procedure

Construction Procedure for Hydroponics Model—Tube Culture

1. Construct a 2′ × 4′ base frame (1″ × 4″ material).
2. Drill 2 1/4″ holes, 1 1/2″ to center from the top edge of the 1″ × 8″ and 1″ × 6″ supports. Space four holes across each member on 6″ centers.
3. Attach supports within the framework 6″ from the upper end and 10″ from the lower end.
4. Cut 2″ PVC pipe into four 3′6″ lengths. Insert into framework and fasten with small wood screws through top of the wooden supports.
5. Drill 2″ holes along the top axis of each pipe, spaced 6″ on center to hold plants.
6. Saw a 2 1/2″ length of PVC pipe in half (end to end). Glue a 2″ cap on one end and install as drain trough at the lower end of the tubes.
7. Place submersible pump in pail, attach tubing and adjust length to reach manifold connection.
8. Attach 1″ manifold along the upper framework. Cap one end and connect to other tubing.
9. Drill manifold and install irrigation header tubes.
10. Friction fit 2″ PVC caps to upper ends of tubes. Drill 1/4″ hole in each and insert header tubes.
11. Apply a suitable finish. Observe all safety precautions when working with electrical equipment in a wet environment.

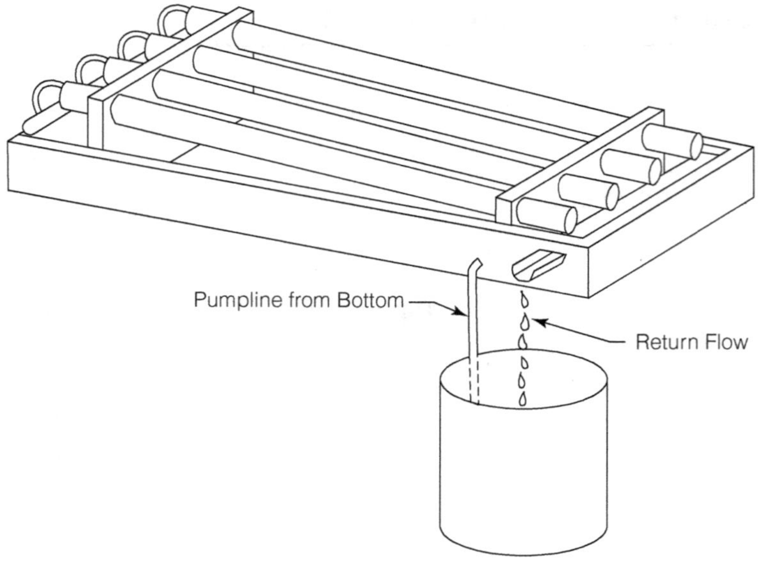

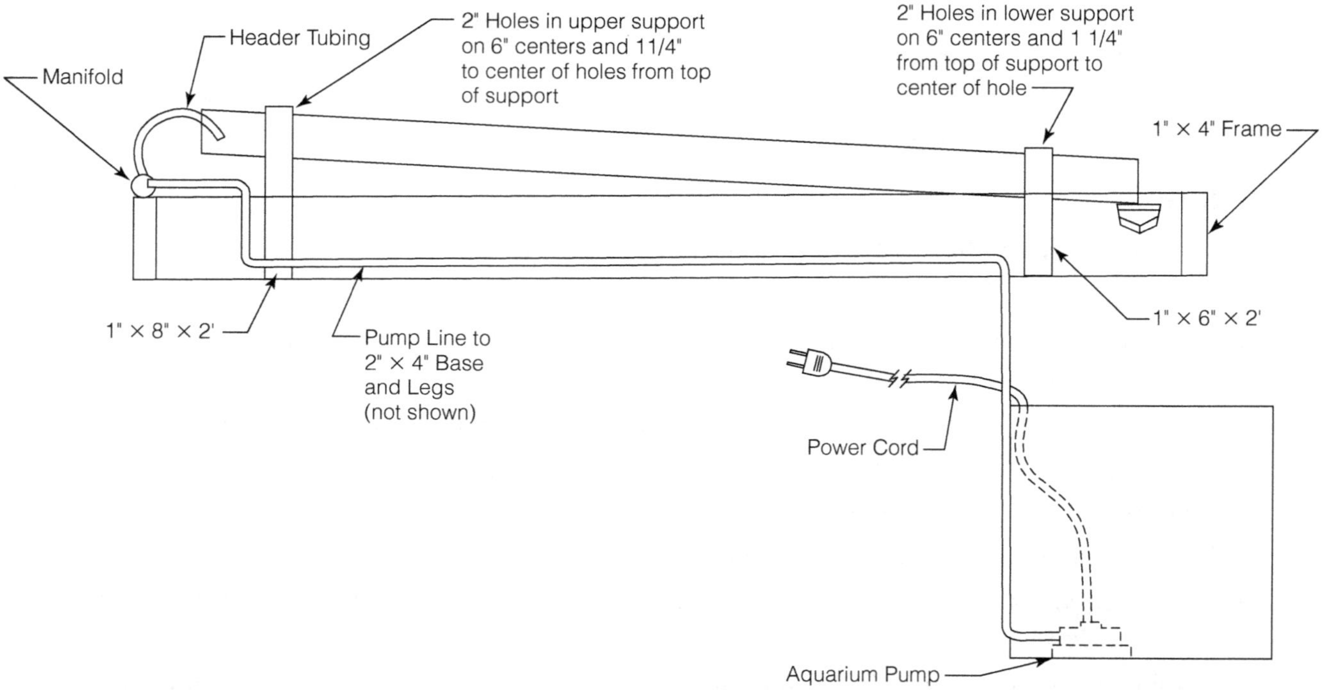

Courtesy University of Maryland at Princess Anne, Dr. Thomas Handwerker

SECTION

3

Effects of Environmental Factors on Plant Growth

Notes

Name: ______________________________ Date: ______________

Exercise 3-1: Calculating Cubic Air Content of a Greenhouse

(by Carroll L. Shry, Jr.)

Purpose

To learn how to calculate the cubic air content of a greenhouse.

Materials

Pen or pencil

Procedure

Study the diagram, formula, and example shown at the right.

1. Using the formula, calculate the cubic air content of the following greenhouses.

	E	R	W	L
A.	8'	14'	25'	100'
B.	6'	12'	35'	200'
C.	9'6"	16'	55'	125'
D.	10'	18'6"	65'	130'
E.	7'	14'	45'	185'

A. __________ cubic content
B. __________ cubic content
C. __________ cubic content
D. __________ cubic content
E. __________ cubic content

2. Measure your own school greenhouse and determine the cubic air volume.

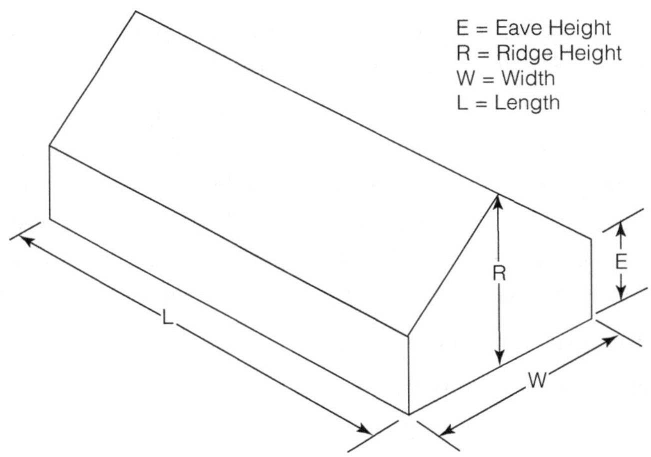

FORMULA: $\frac{E + R}{2} \times W \times L$ = Cubic Content

METHOD: Add E to R and divide by 2. Multiply that answer by W. Multiply this answer by L.

EXAMPLE: E = 6', R = 12', W = 20', L = 100'

$\frac{6 + 12}{2} \times 20 \times 100$ = Cubic Feet

$\frac{18}{2} \times 20 \times 100$ = Cubic Feet

$9 \times 20 \times 100$ = Cubic Feet

$180 \times 100 = 18000$ Cubic Feet = Content

NOTE: Method can be used only on even span greenhouses.

Name: ______________________________ Date: ______________

Exercise 3-2: Controlling Plant Flowering

(by Elizabeth Strum)

Purpose

- Describe photoperiodism based on laboratory experiences.
- Describe which part of the plant senses daylight based on laboratory experiences.

Materials

3 flats
Vermiculite or sand
Seeds for short-day plant
Fluorescent grow lights with timer
Bags or boxes to cover plants

Procedure 1

1. Fill three flats with vermiculite or sand. Individual pots (2" × 2") can also be used.
2. Plant the flats with a suitable short-day plant (see Table 3-2A). Biloxi soybeans work well.
3. For the next six weeks provide a 20-hour photoperiod (4 A.M. to 12 P.M.) and a 4-hour dark period for the plants. Use an automatic timer and fluorescent lights to ensure accurate timing.
4. Adjust the lighting schedule so that the plants are exposed to less than the critical daylength. See Table 3-2B for a list of appropriate daylengths.
5. Leave one flat of plants under 20 hours light/4 hours dark as a control. Expose one flat of plants to the proper short-day lighting schedule for ten days to two weeks. Expose another flat of plants to the same short-day lighting schedule for ten days to two weeks, but expose the plants to 15 to 60 minutes of light in the middle of each dark period. This is fairly simple to do if the first flat is either covered with a box or moved to a dark cabinet each night. Adjust the automatic timer so that it turns on each night for the desired period of time.
6. Return the plants to the regular lighting schedule for four weeks (20 hours light/4 hours dark).
7. Record your observations in Table 3-2C.

Table 3-2A

Flowering Behavior of Some Selected Plants	
Short-day Plants	
Chrysanthemum	*Chrysanthemum morifolium*
Cocklebur	*Xanthium strumarium*
Biloxi soybean	*Glycine max* cv. Biloxi
Japanese morning glory	*Pharbitis nil* strain Violet
Maryland Mammoth tobacco	*Nicotiana tabacum* cv. Maryland Mammoth
Cotton	*Gossypium hirsutum*
Strawberry	*Fragaria spp.*

Name: ______________________________ Date: ______________

Table 3-2A

Flowering Behavior of Some Selected Plants	
Long-day Plants	
Sugar beet	*Beta vulgaris*
Winter barley	*Hordeum vulgare*
Spinach	*Spinacia oleracea*
Oat	*Avena sativa*
Rye-grass	*Lolium temulentum*
Radish	*Raphanus sativus*
Clover	*Trifolium spp.*
Mustard	*Brassica spp.*
Lettuce	*Lactuca sativa* cv. Black-seeded Simpson
Petunia	*Petunia hybrida*
Day-neutral Plants	
Cucumber	*Cucumis sativus*
Corn	*Zea mays*
Tomato	*Lycopersicon sp.*

Table 3-2B

Critical Daylength of Selected Short-Day Plants	
Day Plants	
Maryland Mammoth tobacco	12 hours
Biloxi soybean	14 hours
Japanese morning glory	5 hours
Cocklebur	15.5 hours
Chrysanthemum	13 hours

Table 3-2C

Measurements of Flowering				
Light Regime	**# Plants in Group**	**# Plants Flowering**	**# Flowers on Flowering Plants**	**% of Flowering Plants Out of Total**
Control				
Short-Day Light				
SD Light, Interrupted Night				

Name: ______________________________ Date: ______________

Procedure 2

Use 24 short-day plants (SDP) grown for 6 weeks under 20 hours light/4 hours dark.

1. Leave 6 plants as controls (20 hours light/4 hours dark).
2. Cover the tops only of 6 plants with either a dark plastic bag or a paper bag tied or taped closely to the stem (without damaging it).
3. Cover just the leaves (leave the top of the plant exposed) of 6 plants with a dark plastic or paper bag or cardboard box cut in half vertically and taped around the plant.
4. Cover both the tops and leaves of 6 entire plants (exposed to short-day light).
5. Place all experimental group plants under appropriate SD lighting (see Table 3-2B) for two weeks.
6. Return the experimental groups to regular lighting for 4 weeks or until flowering begins.
7. Record your observations in Table 3-2D.

Table 3-2D

Determining Which Part of the Plant Senses Daylight				
Exposure Regime	# Plants in Group	# Plants Flowering	# Flowers on Flowering Plants	% of Flowering Plants Out of Total
Control plants (regular light)				
Plants with covered tops (SD light)				
Plants with covered leaves (SD light)				
Control plants (SD light)				

Questions for Thought

Photoperiods

1. What percent of the plants in the control group were flowering?

2. What percent of the plants exposed to short daylight were flowering?

Name: ______________________________ Date: ______________

3. Why do you think there is a difference between the two groups?

4. What percent of the plants in the interrupted night group were flowering?

5. Why do you think they were or were not flowering?

What Part of the Plant Senses Daylight?

1. How many plants in the covered-tops group were flowering?

2. How many plants in the covered-leaves group were flowering?

Name: ______________________________ Date: ______________

3. Why do you think there is a difference between these two groups?

4. Can you tell which part of the plant senses the daylength?

Name: ____________________ Date: ____________

Exercise 3-3: Calculating Greenhouse Heating Requirements

(by Steven E. Newman)

Purpose

- To learn how to calculate the heat loss from a greenhouse structure.
- To learn how to determine the size and capacity of a heating system to maintain a desired greenhouse temperature.

Materials

Pencil or pen
Paper

Background

When constructing a greenhouse, one of the most important things to consider is the size of the heating plant or heating components. The same process must be undertaken regardless of whether the greenhouse is to be heated with a boiler or with unit heaters. Several factors must be included in the calculations used to determine the heat requirements of a greenhouse: surface area exposed to the elements; glazing material; the physical condition and its structural components; the average wind speed during the heating months; and the temperature differential, that is, the difference between the outside design temperature and the inside design temperature. The outside design temperature is the lowest average expected temperature during the heating season for the local area. The inside design temperature is often considered to be the lowest acceptable temperature tolerated for any given crop. The heat requirements of a greenhouse can be determined using the formula:

$$H = Ka\ (t_1 - t_2)$$

where H = the heat required, in Btu per hour
K = the heat transfer coefficient for the glazing material used
a = the exposed surface area, in square feet
t_1 = desired inside temperature, in °F
t_2 = outside temperature, in °F

This formula accounts for the heat lost through the covering of the house. It does not take into account heat loss due to high winds, loose glass, or winter ventilation (when needed). To allow for these losses, an additional 10 percent can be added to the heat requirements.

Name: ______________________________ Date: ______________

Procedure 1

1. Calculate the surface area of a tight-metal-frame glass greenhouse with a concrete-block knee wall. Consider that the average minimum temperature is 10°F with a design temperature of 70°F and a typical wind speed of 20 mph.

Greenhouse Surface	Formula	Answer	Two Surfaces	Total Square Feet
Area of end wall A	= 8 ft × 60 ft	= 480 ft^2	× 2 =	960 ft^2
Area of side wall B	= 8 ft × 200 ft	= 1,600 ft^2	× 2 =	3,200 ft^2
Area of gable C	= 2 × ($^1/_2$ × 15 ft × 30 ft)*	= 450 ft^2	× 2 =	900 ft^2
Area of roof D	= 33.5 ft × 200 ft**	= 6,700 ft^2	× 2 =	13,400 ft^2
Area of end knee wall E	= 2 ft × 60 ft	= 120 ft^2	× 2 =	240 ft^2
Area of side knee wall F	= 2 ft × 200 ft	= 400 ft^2	× 2 =	800 ft^2
			Total surface area	19,500 ft^2

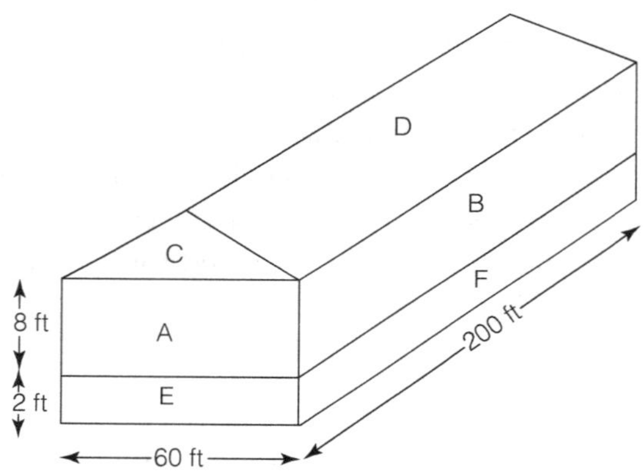

* The gable is essentially two right triangles butted up to each other. They are equal in size when the greenhouse is an even-span structure. Most greenhouses have a roof pitch that is 6 by 12, which means that for every 12 units of width, the roof rises 6 units, making a triangle where the base (b) is 12 units and the height (h) is 6 units. Remember that the area of a triangle is ½ the base times the height.
In our example the base is 30 ft (½ the width of 60 ft) and the height is 15 ft (a 6:12 pitch converts to 15:30) or ½ × 15 ft × 30 ft = 225 ft^2 for half of the gable or 450 ft^2 for the full gable.

**To determine the surface area of one section of the roof, multiply the length times the width. To calculate the width, use the formula $a^2 + b^2 = c^2$, where a = the rise or 15 ft, b = ½ of the gable width or 30 ft, and c = the width of the roof section.

$$15^2 + 30^2 = c^2 \text{ or } 225 + 900 = 1{,}125 = c^2$$
$$c = \sqrt{1{,}125} = 33.5 \text{ ft}$$

Name: ______________________________ Date: ______________

2. Determine the K-value or the heat transfer coefficient. The heat transfer coefficient is that numerical value given to the relative amount of heat transmitted or transferred through a material. The units are typically given as Btu · hr^{-1} · ft^{-2} · $°F^{-1}$.

Note: Some refer to the K-value as the U-value, and it is the inverse of the R-value commonly used for insulation products ($U = \frac{1}{R}$).

Materials*	K-value ($Btu \cdot hr^{-1} \cdot ft^{-2} \cdot °F^{-1}$)
Glass, single	1.13
Glass, double glazing	0.70
Single-film plastic	1.20
Double-film plastic (inflated)	0.70
Corrugated FRP panels	1.20
Corrugated polycarbonate	1.20
Plastic structured sheet (winter)	
16 mm thick	0.58
8 mm thick	0.65
6 mm thick	0.72
Concrete block (knee walls)	0.51
* National Greenhouse Manufacturer's Association	

3. Determine the desired inside temperature in °F (t_1) and the outside temperature in °F (t_2). Typically, when determining the inside and outside temperatures one must consider the crop to be grown and the average minimum temperature of the greenhouse location. There are many tools that a greenhouse manager can use to determine the average minimum temperature. The figure below is average extreme minimum temperature for the United States from the National Oceanic and Atmospheric Administration (NOAA).

Extreme Minimum Temperature (°F)
Jan-Dec 2006

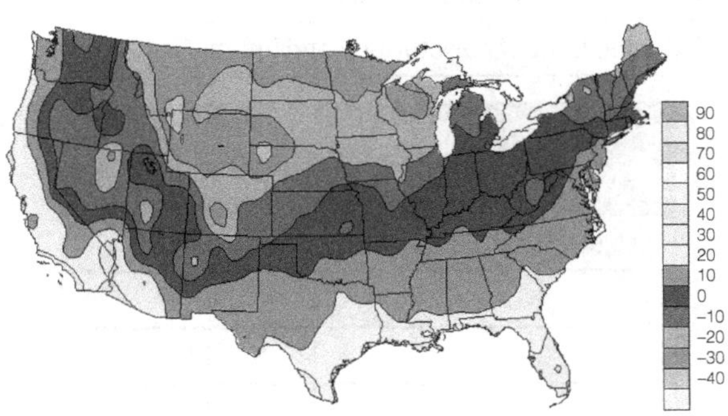

Climate Prediction Center, NOAA
Computer generated contours
Based on preliminary data

For our example, we are using an inside temperature of 70°F and an outside temperature of 10°F. Some call this the ΔT ($t_1 - t_2$), which equals 60°F.

Name: ______________________________ Date: ______________

4. Apply the formula $H = Ka\,(t_1 - t_2)$ using the table below.

Greenhouse Surface	Surface Area (ft^2)	K-value ($Btu \cdot hr^{-1} \cdot ft^{-2} \cdot °F^{-1}$)	$\Delta T\ (t_1 - t_2)$ (°F)	Heat Loss ($Btu \cdot hr^{-1}$)
End walls	960 ft^2	1.13	60°F	65,088
Side walls	3,200 ft^2	1.13	60°F	216,960
Gables	900 ft^2	1.13	60°F	61,020
Roof	13,400 ft^2	1.13	60°F	908,520
Knee walls (end)	240 ft^2	0.51	60°F	7,344
Knee walls (side)	800 ft^2	0.51	60°F	24,480
				H = 1,283,412

$$H = 1{,}283{,}412\ Btu \cdot hr^{-1}$$

5. Apply correction factors for house construction and for wind speed. The house construction factor (m) refers to how well the greenhouse glazing is attached. Loosely attached glazing increases the number of air exchanges per hour. The correction factor for wind speed (W) refers to how much the energy exchange of a greenhouse is increased when the wind is blowing. Consider the following tables.

Air Exchanges for Various Greenhouse Construction Types (M)	
Greenhouse Construction	**Air Exchanges Per Hour**
Metal with glass	1.08
Wood and steel greenhouse with glass	1.05
Wood greenhouse with glass and tight	1.00
Wood greenhouse with glass and moderately tight	1.13
Wood greenhouse with glass and loose	1.25
Wood greenhouse with corrugated plastic	0.95
Metal greenhouse with corrugated plastic	1.00
Metal greenhouse with double glass	0.70
Metal greenhouse with single layer of polyethylene film	1.00
Metal greenhouse with double layer of polyethylene and inflated	0.70

Correction Factors for Wind Speed (W)	
Wind Speed (mph)	**Correction Factor**
Less than or equal to 15	1.00
20	1.04
25	1.08
30	1.12
35	1.16

$$H_{corrected} = H \times M \times W$$
$$H_{corrected} = 1{,}283{,}412\ Btu \cdot hr^{-1} \times 1.08 \times 1.04$$
$$H_{corrected} = 1{,}441{,}528.4\ Btu \cdot hr^{-1}$$

Name: ______________________________ Date: ______________

Complete the Following Exercises

1. Use the previous greenhouse example, but change the glazing to 16-mm-thick plastic structured sheet and calculate the $H_{corrected}$ value.

Greenhouse Surface	Surface Area (ft^2)	K-value ($Btu \cdot hr^{-1} \cdot ft^{-2} \cdot {}^{\circ}F^{-1}$)	ΔT ($t_1 - t_2$) (°F)	Heat Loss ($Btu \cdot hr^{-1}$)
End walls				
Side walls				
Gables				
Roof				
Knee walls (end)				
Knee walls (side)				

$H_{corrected} = H \times M \times W$

$H_{corrected}$ = ____________ $Btu \cdot hr^{-1}$ × ________ × ________

$H_{corrected}$ = ____________ $Btu \cdot hr^{-1}$

2. How is the value of $H_{corrected}$ value for the 16-mm-thick plastic structured sheet different from the previous example?

____

3. Calculate the $H_{corrected}$ value for a tight-metal-frame greenhouse with corrugated plastic glazing and no knee wall with 8-foot sidewalls. The width of the greenhouse is 35 feet and the length is 100 feet. The gable is a traditional 6:12 pitch. Consider that the average minimum temperature is –10°F with a design temperature of 60°F and a typical wind speed of 30 mph.

Greenhouse Surface	Surface Area (ft^2)	K-value ($Btu \cdot hr^{-1} \cdot ft^{-2} \cdot {}^{\circ}F^{-1}$)	ΔT ($t_1 - t_2$) (°F)	Heat Loss ($Btu \cdot hr^{-1}$)
End walls				
Side walls				
Gables				
Roof				
Knee walls (end)				
Knee walls (side)				

$H_{corrected} = H \times M \times W$

$H_{corrected}$ = ____________ $Btu \cdot hr^{-1}$ × ________ × ________

$H_{corrected}$ = ____________ $Btu \cdot hr^{-1}$

Name: ______________________________ Date: ______________

4. Calculate the $H_{corrected}$ value for a range of four (4) gutter-connected tight-metal-frame greenhouses with corrugated plastic glazing and no knee wall with 8-foot sidewalls. The width of each section of greenhouse is 35 feet and the length is 100 feet. The gable is a traditional 6:12 pitch. Consider that the average minimum temperature is –10°F with a design temperature of 60°F and a typical wind speed of 30 mph.

Greenhouse Surface	Surface Area (ft^2)	K-value ($Btu \cdot hr^{-1} \cdot ft^{-2} \cdot °F^{-1}$)	ΔT ($t_1 - t_2$) (°F)	Heat Loss ($Btu \cdot hr^{-1}$)
End walls				
Side walls				
Gables				
Roof				
Knee walls (end)				
Knee walls (side)				

$$H_{corrected} = H \times M \times W$$

$H_{corrected}$ = ______________ $Btu \cdot hr^{-1}$ × ________ × ________

$H_{corrected}$ = ______________ $Btu \cdot hr^{-1}$

5. a. How is the $H_{corrected}$ value for the single greenhouse in exercise 3 different as compared to the $H_{corrected}$ value determined for exercise 4?

____

____

b. How would the $H_{corrected}$ value be different if the four greenhouse structures were constructed as standalone buildings and not gutter connected?

____

____

____

Bonus question:

Calculate the $H_{corrected}$ value for a metal-frame Quonset greenhouse roof with inflated double-layer polyethylene and no knee wall. The width of the greenhouse is 35 feet and the length is 100 feet. Consider that the greenhouse will be in Columbus, Ohio, with a design temperature of 50°F and a typical wind speed of 25 mph.

SECTION

4

Growing Media

Notes

Name: ______________________________ Date: ______________

Exercise 4-1: Physical Properties of Soils

(by Catherine A. Teare Ketter)

Purpose

To distinguish the physical and chemical properties of organic and inorganic soil layers that define soil type and determine suitability for agricultural practices. Specifically, you should

- Distinguish between the organic and inorganic components of soils.
- Determine the pH of the soil and relate pH to the soil's parent material.
- Distinguish between gravel, sand, loam, silt, and clay and relate these terms to relative particle sizes in a soil sample.
- Relate water saturation rates to soil texture.
- Make recommendations about soil tillage based upon the soil's physical characteristics, slope, and climate.

Materials

Procedure 1

Air-dried soil from samples A, B, and C
25-ml beaker
pH meter or paper

Procedure 2

2.00 mm, 0.5 mm, and 0.002 mm sieves
Pan
12 small paper bags
Oven-dried soil from samples A, B, and C
Scale

Procedure 3

Soil from samples A, B, and C
3 canvas bags
Scale

Procedure 1—Measuring Soil pH

Since soil pH is one of the physical properties that determines how well plants grow, it is important to be able to measure soil pH. The method you will use involves measuring the pH of soil suspended in distilled water. Your teacher will give you three soil samples: Sample A—cultivated field with nitrogen fertilizer added recently; Sample B—forest soil; and Sample C—soil from an undisturbed fencerow or pasture.

Hypothesis

Write a hypothesis that relates soil pH to the type of cultivation/vegetation in each sample site.

Name: ______________________________ Date: ______________

Which site do you anticipate will have the lowest pH? Why?

Which site do you think will have the most alkaline pH? Why?

Procedure for Measuring Soil Water pH

1. Weigh out 10 g of air-dried soil from Sample A.
2. Rinse a 25-ml beaker with distilled water and dry.
3. Place your 10-g soil sample in the beaker and add 10 ml of distilled water to the soil.
4. Mix your soil and water in the beaker for 5 seconds.
5. Let the mixture stand for 10 minutes.
6. Gently swirl the soil-water mixture and lower the pH electrode so that it is just above the soil particles. If you are using pH paper, be certain that your paper is at this point in the beaker.
7. Allow the pH meter to come to equilibrium and take a reading.
8. Rinse the pH electrode with distilled water.
9. Repeat the procedure with Samples B and C and record the results in the following table.

Table 4-1A

Results of Soil-Water pH for Three Different Samples			
Sample/Site	Sample A Cultivated Field with Fertilizer	Sample B Forest Soil	Sample C Undisturbed Fencerow or Pasture
Sample pH			

Did your results support your hypothesis?

Name: ____________________ Date: ____________

Based upon the results in Table 4-1A and what you know about soils, how do you think soil pH is related to vegetation type and/or agricultural practices?

Procedure 2—Determining Soil Type as a Function of Soil Texture

As previously stated, there are several different particle size classes: gravel, sand, silt, and clay. These particle size classes are based upon the diameter of the particle. The relative proportion of particle sizes in soil gives the soil many of its physical characteristics such as texture, pore size, water infiltration rate, water-holding capacity, and cation-exchange capacity, for example. Using a scale and a set of sieves with different pore sizes, soil type (as a function of soil texture classes—see Figure 4-1) can be determined. Your teacher will give you three soil samples: Sample A—cultivated field with nitrogen fertilizer added recently; Sample B—forest soil; and Sample C—soil from an undisturbed fencerow or pasture. These are the same sites that were used for soil pH.

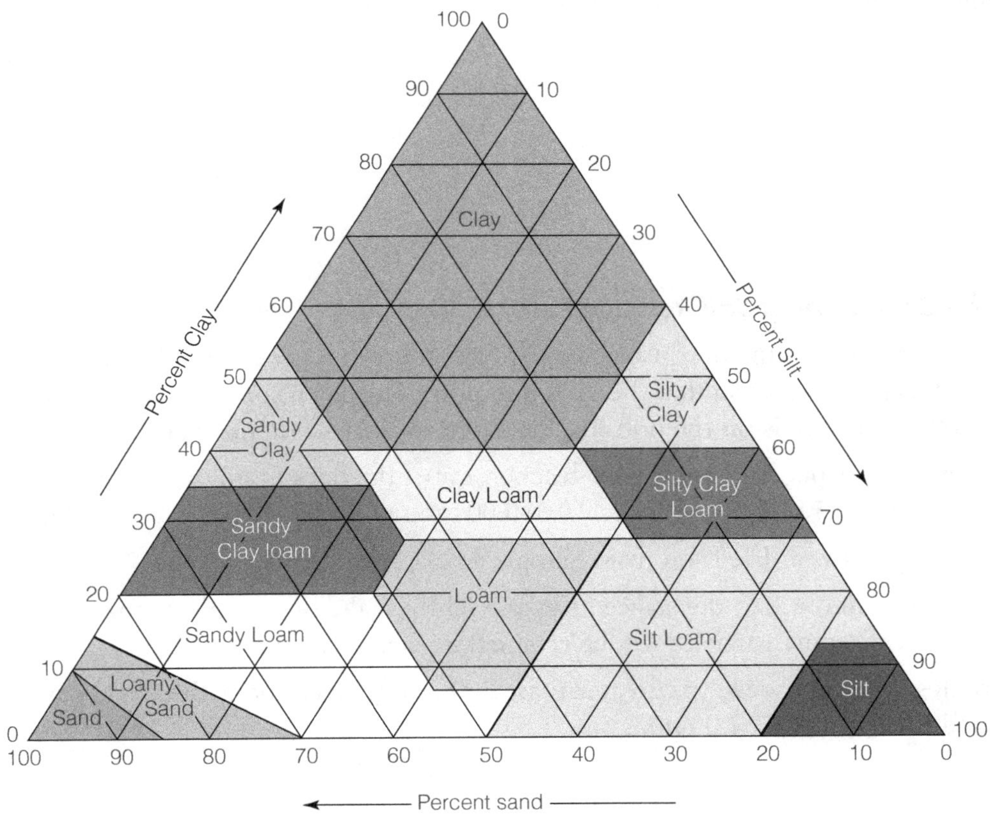

Figure 4-1 Relationship between soil particle size distribution and soil texture class.

Name: ______________________________ Date: ______________

Hypothesis

Write a hypothesis that relates the soil texture to the type of vegetation or agricultural practice used.

Which sample do you expect to have the highest proportion of clay? Why?

Which sample do you expect to have the highest proportion of sand? Why?

Procedure for Determining Soil Relative Particle Size

1. Obtain a set of sieves from your instructor with the following pore sizes: 2.00 mm, 0.5 mm, and 0.002 mm. Stack the sieves so that the 2.00 mm pore sieve is on top, the 0.05 mm sieve is next, and the 0.002 mm sieve is on the bottom. Place the stacked sieves in a pan.
2. Obtain four small paper bags from your teacher. Label the bags as follows: SAMPLE A, > 2.00 mm; SAMPLE A, 2.00–0.5 mm; SAMPLE A, 0.5–0.002 mm; and SAMPLE A, < 0.002 mm.
3. Weigh out 10 g of oven-dried soil from Sample A.
4. Place soil from Sample A in the stacked sieves. Gently shake the sample so that the soil falls through the sieve (some particles will be retained in each screen).
5. The particles in the top sieve are greater than 2.00 mm (gravel); pour these soil particles into the bag labeled SAMPLE A, > 2.00 mm.

Name: ______________________ Date: ______________

6. The particles in the next sieve are smaller than 2.00 mm in diameter and larger than 0.5 mm (sand). Pour these particles into the bag labeled SAMPLE A, 2.00–0.5 mm.
7. The particles in the next sieve are smaller than 0.5 mm in diameter and larger than 0.002 mm (silt). Pour these particles into the bag labeled SAMPLE A, 0.5–0.002 mm.
8. The particles in the bottom of the pan are smaller than 0.002 mm in diameter and represent the clay fraction of the soil. Pour these particles into the bag labeled SAMPLE A, < 0.002 mm.
9. Weigh the soil in each of your four paper bags and record their weights (g) in Table 4-1B.
10. Repeat the procedures for soil from Samples B and C. Remember to use new paper bags each time to get the most accurate weight estimates. Record the weights (g) of each texture class in Table 4-1B.

Table 4-1B

Soil Fraction Weights for Three Different Soil Samples			
Weight of Soil Fraction (g)	**Sample A Cultivated Field with Fertilizer**	**Sample B Forest Soil**	**Sample C Undisturbed Fencerow or Pasture**
Gravel (> 2.00 mm)			
Sand (2.00-0.5 mm)			
Silt (0.5-0.002 mm)			
Clay (< 0.002 mm)			

Did your results support your hypothesis?

Calculate the percentage of each size class for each sample.

$$\% = \frac{\text{Weight of soil fraction (g)}}{\text{Total weight of sample (10 g)}}$$

See Figure 4-1, soil texture triangle. What type of soil is found in Sample A?

Name: ______________________________ Date: ______________

What type of soil is found in Sample B?

What kind of soil is found in Sample C?

Given your understanding of the relationship between soil pH and cation-exchange capacity, do you think that the relative proportion of gravel, sand, silt, and clay in the soil has an effect on soil water pH? Relate your answer to the data from Table 4-1A.

Procedure 3—Estimating Water-holding Capacity for Different Soils

Water is essential for all plant growth. Animals also require water for survival. Many animals rely on stored water from plants as their primary water supply. How does the type of soil relate to the amount of water the soil can hold? Is it possible for soil to hold too much water? Soil is made of structural units called peds. Peds determine the structure of soil (in contrast to soil texture, which is related to particle size). Factors which determine the structure of soils are soil texture (relative proportion of different particle sizes), organic matter content, amount of calcium carbonate and/or iron oxides, freeze-thaw cycles, biological activity (impact of earthworms, fungi, and bacteria in soil), and soil cultivation practices. All of these factors contribute to the size and shape of soil aggregates. Soils have spaces between the peds through which water, water-soluble minerals and organic nutrients, and atmospheric gases can move. When these spaces are completely filled with water, the soil is saturated. Eventually, some of this water is replaced by atmospheric gases (oxygen—O_2, carbon dioxide—CO_2, and nitrogen—N_2). Most plants do not thrive in poor drainage conditions resulting in permanently saturated soils. Water-logged soils typically are low in O_2. Only plants and microorganisms adapted to low oxygen conditions survive in such locations. Swamps, bogs, and tidal marshes are examples of areas with permanently saturated soils under

Name: ______________________ Date: ______________

normal conditions. These areas do not make good sites for the cultivation of most plant crops. The amount of water which soils can hold is dependent upon the ped size, soil texture, and slope of the site. Sites with moderate to steep slopes drain water faster than flat areas. Using soil samples from three sites: a cultivated field, a forest, and a fencerow or pasture, you will estimate the amount of water in a saturated sample.

Table 4-1C

Water Saturation Estimates for Three Different Soil Samples			
Sample Site	**Wet Weight**	**Dry Weight**	**Water (g) = Wet Weight – Dry Weight**
Sample A Cultivated field with fertilizer			
Sample B Forest soil			
Sample C Undisturbed fencerow or pasture			

Hypothesis

You have classified the soil types for each of these three sites based upon soil texture. Write a hypothesis regarding the relationship between soil type and the amount of water necessary to saturate the soil.

Which site do you think will require the most water for saturation? Why?

Procedure for Measuring Soil Saturation

1. Measure 100 ml of soil from each of the three sites: A, B, and C.
2. Place each sample in a canvas bag, tie the bag tightly and weigh the sample. Record the “dry” weight in Table 4-1C.
3. Thoroughly saturate each bag with water by placing it underneath running water in the sink or putting the bag in a pan of water and allowing the bag to absorb as much as it can (approximately 20 to 30 minutes).

Name: ______________________________ Date: ______________

4. When the samples are saturated, allow the excess water to drain.
5. Weigh each of the samples again and record the "wet" weight in Table 4-1C.
6. Calculate the weight of the water by subtracting the dry weight from the wet weight and record the amount in Table 4-1C.

Did your results support your conclusion?

How do you think vegetational cover affects soil saturation?

Summary

Given the measurements of soil water pH, soil texture, and water saturation for each of these three sites, assuming that all other factors are the same (aspect, slope, climate), which of the three sites is best suited for cultivation? Include in your answer all of the evidence which you have gathered in the three experiments on physical soil characteristics.

Has the type of vegetation found in each area affected the physical properties of the soils? How?

Name: ______________________________ Date: ______________

In which area would you expect to find the greatest diversity of living organisms in the soil? Why?

What recommendations could you make for producers who are interested in maintaining healthy soil based on these experiments?

Name: ______________________________ Date: ______________

Exercise 4-2: Biological Properties of Soils

(by Catherine A. Teare Kitter)

Purpose

To investigate the biological properties and roles of soil. Specifically, you should

- Identify the portion of the soil horizon containing the greatest concentration of microorganisms.
- Define the role of algae, bacteria, and fungi in soils.
- Distinguish between algal, bacterial, and fungal cells using a compound light microscope.
- Describe the impact of agricultural practices on soil microbes.
- Observe *Rhizobium* in the root nodules of a legume.
- Estimate population numbers from a given soil sample using serial dilution microbial culture techniques.
- Describe the role of soil macroorganisms such as earthworms, nematodes, and anthropods.
- Observe soil macroorganisms collected in pitfall traps.

Materials

Procedure 1

Soil from samples A, B, and C
Soil-sterile water suspension
Sterile pipette
Microscope
Slides and coverslips

Procedure 2

Legume root
Crystal violet stain

Procedure 3

Sterile flask
Distilled water
3 test tubes
6 petri dishes
Flask of nutrient agar

Procedure 4

Pitfall trap samples from sites A, B, and C
Dissecting microscope or magnifying lens

Procedure 1—Observing Soil Microorganisms

The presence of soil microorganisms provides an indication of soil quality. Healthy soils have an abundance of soil bacteria, fungi, and algae. You will receive a soil-sterile water suspension from your instructor which you will examine under the light microscope for the presence of soil microorganisms.

Name: ______________________ Date: ______________

Your teacher will provide you with three soil samples: Sample A—field under cultivation with leguminous plants, Sample B—mixed forest soil, and Sample C—soil from a pasture currently in use.

Hypothesis

Write a hypothesis that relates the kinds of soil organisms, bacteria, fungi, and algae, to the type of cultivation or vegetation in each sample site.

1. With the cap secured, shake the soil-sterile water suspension from Sample A. Allow the large particles to fall to the bottom of the jar.
2. Using a sterile pipette, pipette a few drops of the soil-water suspension onto the middle of a flat slide.
3. With the coverslip at an angle, drop the coverslip on the soil-water suspension.
4. If you have air bubbles under your coverslip, gently tap the surface of the cover slip with your pencil eraser or the blunt end of a dissecting needle. The bubbles should move to the outside of the coverslip.
5. Turn on your microscope light and make any necessary adjustments (center the slide on the stage, adjust the light using the diaphragm, etc.).
6. Begin with the low power objective and bring the slide into focus. Bacterial cells are too small to be seen with the low power objective, but fungi and some algae may be visible.
7. Without moving the slide, switch to the medium and high power objectives. **Do not use the coarse adjustment knob with the high power objective!**
8. Algal cells will appear rectangular or circular with a definite outside wall. The cells of soil algae are usually green, blue-green (very small-cell, blue-green algae also called cyanobacteria), or a brownish color (diatoms).

   Do you see any algae present in the soil-water suspension from Sample A? If so, draw what you see in the space provided.

Name: ______________________________ Date: ______________

9. Fungal cells are clear in color, appearing as long, thin threads, hyphae, with definite walls between the cells. Some of the hyphae may have broken when you shook your sample jar.

Do you see any fungal cells in the soil-water suspension from Sample A? If so, draw what you see in the space provided.

10. Bacterial cells are very small (150 to 500 μm). Many are too small for inexperienced eyes to see without the use of the oil immersion objective lens. Your teacher will review the use of the oil immersion objective lens with you. Be certain that the slide is in focus using the high power objective lens, and adjust your condenser so that it is closest to the stage. Increase the amount of light by opening the condenser diaphragm. Move the revolving nosepiece in the direction of the oil immersion lens so that the slide is halfway between the high power objective and the oil immersion objective—DO NOT put the oil immersion objective in place yet. Put 1 or 2 drops of immersion oil on your coverslip and slide the oil immersion lens in place. Wait a minute or two to assure that the oil has covered the oil immersion objective. Using the fine adjustment knob only, bring the slide into focus. Bacterial cells come in three shapes (see Figure 4-2A): spherical (coccus), rod-shaped (bacillus), and spiral (spirilla). You should see single cells with a pronounced cell wall and little color.

Do you see bacteria cells in your soil-water suspension from Sample A? If you see any bacteria cells, label the cells according to shape.

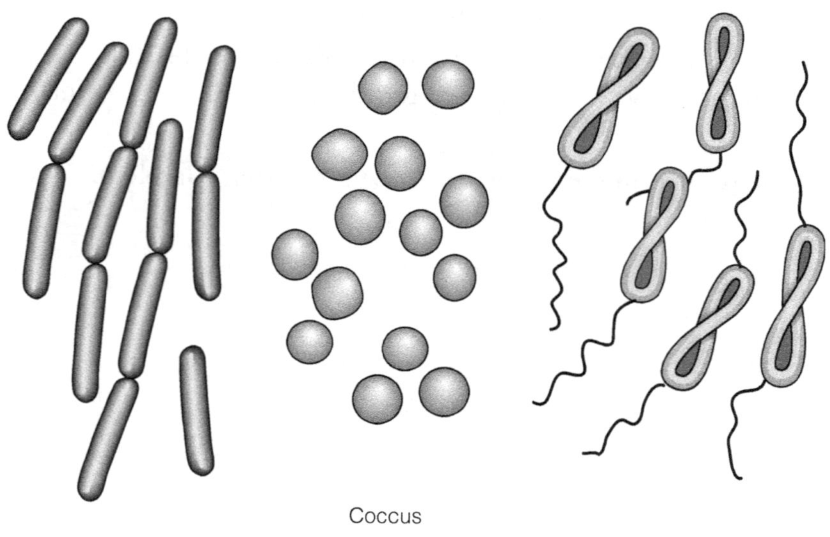

Figure 4-2A Shapes of bacterial cells.

Name: ______________________________ Date: ______________

11. Bacterial cells are easier to see with the aid of special colorizing agents—stains. Move the oil immersion lens out of place carefully and clean the immersion oil off the objective lens using the cleaning solution provided by your teacher and lens paper. Remove your coverslip and put 1 drop of crystal violet stain under your cover slip; and add more water if necessary. Put a new, clean coverslip on your slide and focus your slide using the low, medium, and high power objectives. Use the oil immersion objective as before, and look at your slide.

Are there more bacterial cells visible with the crystal violet stain? How many different shapes of cells do you see?

12. Repeat steps 1 through 11 for Samples B and C.

Were algae present in any of your soil samples? From which site?

What would the presence of algae suggest about the amount of water the soil is holding?

Were fungal hyphae observable in any of your soil samples? From which sample?

Name: ______________________________ Date: ______________

What would fungi indicate about the type(s) of vegetation in the area?

What roles do fungi play in soils?

Were bacteria present in any of your soil-water suspensions? In which sample(s)?

In which sample were the bacteria most prevalent?

In terms of your original hypothesis, how are vegetation cover and tillage practices related to the numbers and kinds of soil microorganisms?

Name: ______________________________ Date: ______________

Procedure 2—Observing Legume Root Nodules

Small amounts of available nitrogen in the soil can limit plant growth. Many important cash crops such as wheat, corn/maize, sorghum, cotton, tobacco, and oats have high nitrogen requirements. The practice of crop rotation, planting different crops in a field (or using the field as a pasture) from one growing season to the next, can greatly improve soil quality and reduce the amount of chemical fertilizer required for profitable yields. Practitioners of crop rotation frequently use leguminous plants as an alternative crop to winter wheat, corn, and cotton. Nitrogen-fixing bacteria belonging to the genus *Rhizobium* are found in association with bumps (nodules) on the roots of peas, beans, clovers, and other leguminous plants.

Hypothesis

In which of the three soil samples (Sample A—field under cultivation with leguminous plants, Sample B—mixed forest soil, and Sample C—soil from a pasture currently in use) would you expect available nitrogen to be greatest?

Explain how the amount of available soil nitrogen is related to the differences in cultivation practices among the three sites.

1. Obtain a legume root with visible nodules from your teacher. Using a dissecting needle, put 1 or 2 nodules on a clean microscope slide.
2. Put several drops of water and 1 drop of crystal violet stain on the nodules.
3. Add a coverslip. Wrap the slide in a paper towel, coverslip side up.
4. Using the eraser end of your pencil or the blunt end of your dissecting needle, crush the nodules until the coverslip is flat.
5. Using the low, medium, and high power objectives focus the slide.
6. Use the procedure for the oil immersion objective lens, and look at the nodules at 1000× magnification.

Name: ______________________________ Date: ______________

Draw what you see in the space provided. What shape are the *Rhizobium* bacteria (coccus, bacillus, or spirillum)?

Procedure 3—Estimating Soil Bacterial Populations

The numbers and kinds of soil bacteria provide agricultural scientists with an indication of soil health and potential productivity. Techniques for estimating soil bacterial populations are useful tools in evaluating the suitability of soils for different agricultural practices. Your class will be divided into smaller groups, each group receiving a soil sample from one of the three sites: Sample A—field under cultivation with leguminous plants, Sample B—mixed forest soil, and Sample C—soil from a pasture currently in use. Using a microbiological lab technique called serial dilution (see Figure 4-2B), your group will estimate the number of bacteria present in 1 g of soil from your site. You will compare the estimates of bacterial population size with the other groups in your class.

Hypothesis

Write a hypothesis that relates the size of soil bacterial populations to the type of cultivation or vegetation in each site.

1. Weigh out 1 g of soil from your sample.
2. Remove the stopper from a sterile flask, add 1 g of soil to 10 ml of sterile distilled water. Do not touch the bottom of the stopper or lay it on your table or lab bench to avoid potential contamination of your soil-sterile water solution. Put the stopper back in place and shake the flask so that the soil is suspended throughout the water column.
3. Allow the large particles to settle to the bottom of the flask.
4. Obtain 3 test tubes, each containing 9 ml of sterile water and label them: "1/10", "1/100", and "1/1000". Make a serial dilution of your original water sample as shown in the diagram. Use the same pipette for all three transfers, but rinse it between each transfer. In other words, after discharging the contents of the pipette into the new dilution tube, rinse the pipette by drawing up some of the medium and emptying the pipette again. Remember to mix each tube before continuing the dilution.

Name: ______________________________ Date: ______________

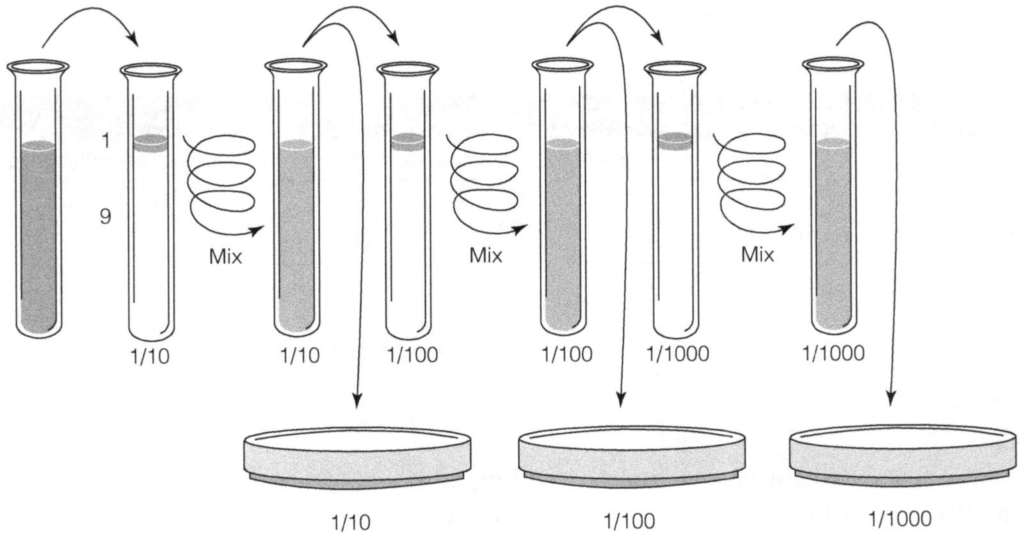

Figure 4-2B Serial dilution technique.

5. Label 6 sterile petri dishes (2 each) "1/10", "1/100", and "1/1000", and add *your initials*, *lab period*, and *type of water used*.
6. Using a clean pipette and beginning with the 1/1000 dilution, inoculate each of the 6 plates with 1 ml from the appropriate dilution tube.
7. Obtain a flask of molten nutrient agar from the heated water bath. Using aseptic technique, pour enough agar into one of the inoculated dishes to cover $^2/_3$; of the plate. Immediately mix the molten agar with the diluted water sample by gently rotating the plate clockwise and counterclockwise without splashing or lifting it from the lab bench. Repeat the process with the other inoculated petri dishes. Allow the plates to cool until the agar is well hardened. Place a piece of tape on either side of each dish to prevent them from coming open. Each person is to take home 3 plates (one plate of each dilution) and maintain them at room temperature. After 48 hours count the colonies present on those plates where counting is feasible.

In the space provided, record the type of water used, the number of colonies present on each plate and calculate the number of viable bacteria present per ml of the original soil-water sample, taking into account the pre-dilution performed on the soil (10 ml of sterile water).

Next week in lab, compare your numbers to those of your lab partner and the other students working on the same site. (Students often wonder why we do not provide a formula for making the calculations. The answer is that by reasoning it out for yourself and coming up with your own formula, you will have a much better understanding of what is involved than you would if you simply plugged your data into a formula we gave you.)

Pre-dilution, if any, for your sample (A, B, and C) 1 g soil/10 ml sterile water

Colonies present per plate:

1/10 ____________________ 1/100 ____________________ 1/1000 ____________________

Viable bacteria present per ml of the original soil sample:

____________________ cells/1 g/ml

Name: ______________________________ Date: ______________

Collect the data from the class and record it in Table 4-2A.

Table 4-2A

Bacterial Population Estimates From Three Different Types of Soils		
Sample Site	**Bacterial Population Estimates**	**Average Population Size**
Sample A Cultivated field with leguminous plants		
Sample B Mixed forest soil		
Sample C Pasture currently in use		

An average is calculated for each sample by adding all of the estimates for the site and dividing the total by the number of estimates. For example, if there are four groups with estimates of 150/ml, 250/ml, 500/ml, and 50/ml, the average for the site/sample would be [(150 + 250 + 500 + 50)/4]. You need to calculate the average population size for each of the three sample sites and record the values in the right-hand column in Table 4-2A.

Was your hypothesis supported by the data the class collected?

Were your results different than expected? If so, give some reasons why the results may have been different than you initially predicted.

Procedure 4—Soil Macroorganisms

Arthropods are invertebrates, animals which lack a backbone. Arthropods have exoskeletons, that is, their skeleton is on the outside of the body. All arthropods also have jointed bodies and appendages. Insects, spiders, scorpions, ticks, and mites are all arthropods. Another common soil invertebrate is the nematode. Nematodes are a type of worm which lacks internal segmentation. In terms of numbers, nematodes are the most abundant invertebrates on earth. Nematodes can be found living independently in the soil, but most frequently, nematodes are found in association with plant roots. Some nematodes can cause extensive crop damage by injuring plant roots—this affects the plant's ability to take up water and nutrients. A nematode infestation can greatly reduce corp yield. Most soil invertebrates are beneficial; they incorporate the remains of plant material and humus through the uppermost portions of the soil profile.

Name: ______________________________ Date: ______________

There are many methods for collecting soil macroorganisms. One technique involves the use of core samples, long tubes of specific diameters sunk to a given soil depth. Core samples are used to look at soil microflora, microscopic invertebrates, bacteria, fungi, and algae. Macroscopic animals can be collected in the field using pitfall traps (see Figure 4-2C). You will be given pitfall trap samples collected from the three sites you have been working with: Site A—a field under cultivation with leguminous plants, Site B—a mixed forest, and Site C—a pasture currently in use. You will use a dissecting microscope or a magnifying lens to categorize and count the invertebrates in your sample.

Hypothesis

Write a hypothesis that relates the number of different kinds of soil invertebrates to the type of cultivation practice in each of your three sites.

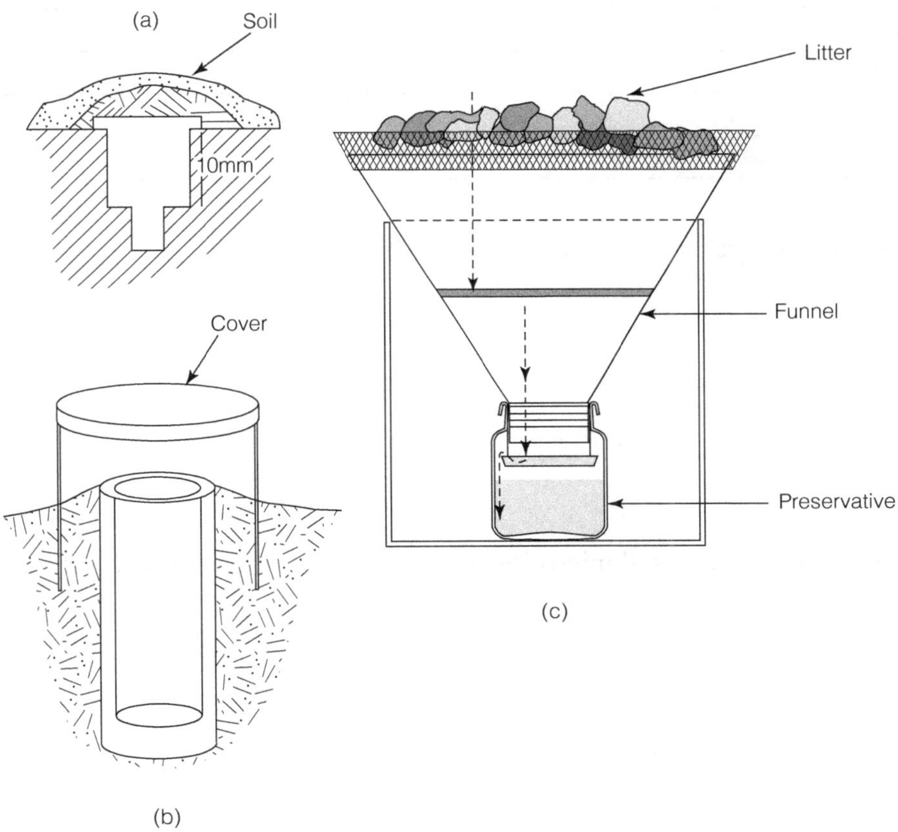

Figure 4-2C Kinds of soil pitfall traps. (a) Soil-covered trap; (b) standards pitfall trap; and (c) litter-covered funnel trap.

1. Empty the organisms from the trap bottle or vial (with alcohol used as a preservative) into the dish provided by your teacher.
2. Place the dish under the dissecting microscope (or alternately, use your hand lens).
3. Separate your animals into groups which look different. Place all of the individual animals belonging to one kind of invertebrate group into smaller bottles with preservative from your teacher. Your teacher will help you to determine which animals are annelids (earthworms),

Name: ______________________________ Date: ______________

nematodes, and arthropods (crustaceans, such as pill bugs); millipedes, centipedes, arachnids (spiders, ticks, and mites), and insects. Record the number of different kinds of soil macroorganisms for each of your three sites in Table 4-2B.

Table 4-2B

Numbers of Different Types of Soil Invertebrates From Three Different Sites			
Type of Invertebrate	**Site A** **Field Under Cultivation with Leguminous Plants**	**Site B** **Mixed Forest**	**Site C** **Pasture Currently in Use**
Nematodes			
Annelids			
Arthropods: Crustaceans			
Arthropods: Millipedes			
Arthropods: Centipedes			
Arthropods: Arachnids			
Arthropods: Insects			

Did your observations support your original hypothesis?

How do you think the presence of soil microorganisms influences the numbers and kinds of soil macroorganisms?

Name: ______________________ Date: ______________

Exercise 4-3: Determining Soil Particle Size

(by Carroll L. Shry, Jr.)

Purpose

To determine the different soil particle sizes in a sample of soil, which influence the movement of air and water in the soil

Materials

Wide mouth pint jars
Soil samples
Liquid detergent

Procedure

1. Select a wide mouth canning jar (as shown to the right) and fill half-full with soil. Do not compact the soil in the jar.

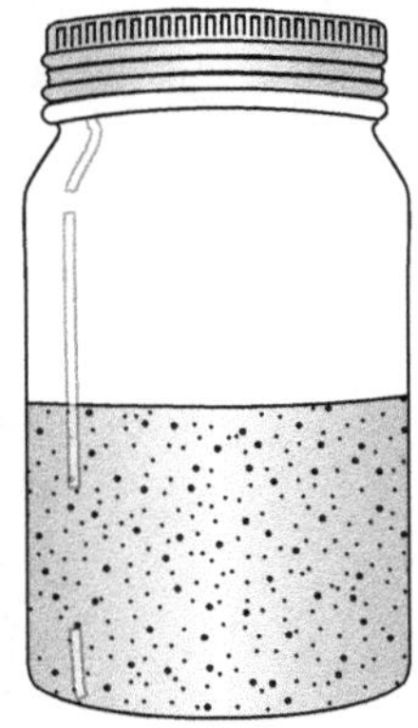

2. Add water until the jar is two-thirds fill and add two drops of liquid detergent, which serves as a wetting agent.

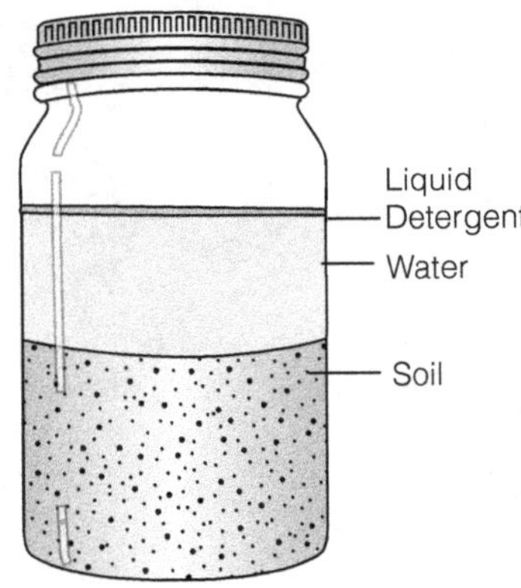

3. Tighten the lid and shake the sample, mixing the soil, water, and wetting agent. Mix the sample thoroughly for 5 minutes. Allow the sample to stand to settle the particles. Observe the sample as it settles at 5 minutes, 10 minutes, 20 minutes, 2 hours, and 24 hours.

   What results did you observe?

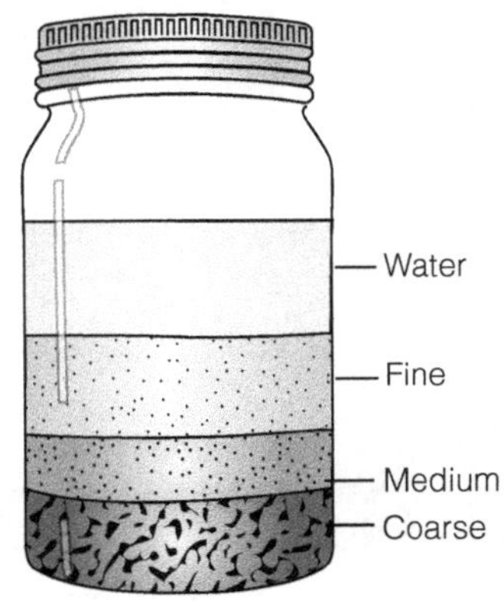

SECTION

5

Nutrition and Watering

Notes

Name: ______________________________ Date: ______________

Exercise 5-1: Role of Minerals in Plant Growth

(by Catherine A. Teare Ketter)

Purpose

To investigate and define the role of minerals in plant growth. Specifically, you should

- Define essential element.
- Define mineral nutrient.
- Distinguish between:
  - — a macroelement and a microelement
  - — a macroelement and a macronutrient
  - — a microelement and micronutrient
- Describe the role of the following essential elements in normal plant functioning:
  - — calcium (Ca)
  - — magnesium (Mg)
  - — nitrogen (N)
  - — phosphorus (P)
  - — potassium (K)
  - — sulfur (S)
- List the deficiency symptoms exhibited by annuals for each of the essential elements.
- Construct and test a hypothesis relating to the quality and quantity of growth in a nutrient solution lacking an essential element.

Table 5-1A

Essential Elements Required for the Growth of Vascular Plants	
Element	**Chemical Symbol**
Macroelements	
Carbon	C
Hydrogen	H
Oxygen	O
Nitrogen	N
Phosphorus	P
Sulfur	S
Potassium	K
Calcium	Ca
Magnesium	Mg
Microelements	
Boron	B
Chlorine	Cl
Copper	Cu
Iron	Fe
Manganese	Mn
Molybdenum	Mo
Zinc	Zn

Name: ______________________________ Date: ______________

Materials

Procedure 1

Young tomato plants
Flats with capillary matting
Marking pen
Nutrient solutions

Procedure 2

Looseleaf notebook

Procedure 1—Mineral Deficiencies

Your instructor will provide you with young tomato plants of a uniform age. Each plant should have at least eight true leaves. Gently shake off as much soil as possible and transplant the tomatoes into the pots provided using sterile vermiculite as a potting medium. Your instructor will divide your class into groups of three or four students. Each group of students will have eight plants: Control, Ca-calcium, N-nitrogen, K-potassium, P-phosphorus, Mg-magnesium, S-sulfur, and Fe-iron. You will water your plants daily with one of eight nutrient solutions. The control solution has all of the necessary macronutrients. A test solution missing one of the essential macronutrients will be applied as needed to one of the tomato plants in each of the seven treatment groups. You will be given seven deficient solutions, each missing one essential macronutrient. Using a permanent marking pen, write the names of the people in your group, the date, and the treatment type (Control, Ca, N, K, P, Mg, S, or Fe) on the plant label provided. When you are finished, your instructor will provide you with a place for your plants. All of the control plants will be placed in flats with capillary matting, and each of the seven treatment groups of plants will be in the same flat with capillary matting. The capillary mats will allow the plants to continually absorb the mineral solutions. You will make and record observations at regular intervals for ten weeks. At the end of the experiment, all of the data collected by each group will be pooled. Your instructor will help you to summarize the results of the experiment.

The symptoms produced by plant nutrient deficiencies relate to the role of the nutrient plays in cellular metabolism. The visible symptoms of several macronutrient deficiencies are summarized below.

Calcium	Symptoms appear at growth points of the roots and stems, flower and fruit set reduced or lacking; root system poorly developed—may be jellylike; leaves yellow, rolled, and curled.
Iron	Effects of iron deficiency systemic; leaf tissue between the veins white and chlorotic—appearing first in youngest leaves; leaf margins and tips scorched, eventual chlorosis/bleaching of all shoots and leaves.
Magnesium	Brittle leaves frequent; green veins with yellow to white leaf tissue between the veins presenting a mottled appearance; leaf drop may occur.
Nitrogen	Shoots spindly—short and thin; growth in young plants stunted; young leaves yellow in appearance, older leaves will be green then turn yellow; leaf drop common.
Phosphorus	Purple accessory pigments (anthocyanins) produce a blue-purple color in the veins; plants may be dwarfed; young leaves blue-green to purple in color.
Potassium	Dead (necrotic) spots appear first on old leaves; irregular leaf surface (crinkled, wrinkled) between the leaf veins; leaves blue-green to purple in color; leaf margins may become bleached (white) or chlorotic (yellow) in appearance.
Sulfur	Stems may be fine/slight; leaf color changing from green to light green to yellow; color change first appears around leaf veins.

Name: ______________________________ Date: ______________

Hypothesis

Which mineral deficient solution do you expect will produce symptoms first? Why?

Write a hypothesis that relates amount of plant growth and/or appearance to each specific macronutrient deficiency.

1. Calcium

2. Nitrogen

3. Potassium

4. Phosphorus

Name: ______________________________ Date: ______________

5. Magnesium

6. Sulfur

7. Iron

Do you expect to see symptoms in the control plant? Explain your answer.

Procedure 2—Data Collection

You will need to use a looseleaf notebook to organize your data. Your instructor will provide you with data sheets for you to make your observations. You will have thirteen sheets for your group. The first sheet will be used to make observations at the time of transplanting (before the test solutions are applied). You will use an additional data sheet each week for ten weeks to record your weekly observations. One data sheet will be used to summarize your group's findings at the end of ten weeks, and the last data sheet will be used to record and summarize class data.

When calculating your summary data, you need to compare your final data table (after ten weeks) to your initial data table. You would record a change in average internode distance (final to initial). For example, if you determined that your average internode distance for the control plant was 2.5 cm and the final value was computed at 6.5 cm, then the change in average stem internode distance was 6.5–2.5 cm or 4.0 cm.

Name: ______________________________ Date: ______________

Table 5-1B

Data Table for Student Group						
Treatment	Plant Height (cm)	Average Stem Internode Distance (cm)	Number of Leaves	Leaf Appearance	Number of Flowers/Fruit	Stem Diameter (cm) at First Pair of Leaves
Control						
Calcium						
Nitrogen						
Potassium						
Phosphorus						
Magnesium						
Sulfur						
Iron						

Table 5-1C

Summary Table for Student Group						
Treatment	Change in Plant Height (cm)	Change in Average Stem Internode Distance (cm)	Change in Number of Leaves (+/−)	Leaf Appearance	Number of Flowers/Fruit	Change in Stem Diameter (cm) at First Pair of Leaves
Control						
Calcium						
Nitrogen						
Potassium						
Phosphorus						
Magnesium						
Sulfur						
Iron						

Table 5-1D

Summary Table for Class Data						
Treatment	Change in Plant Height (cm)	Change in Average Stem Internode Distance (cm)	Change in Number of Leaves (+/−)	Leaf Appearance	Number of Flowers/Fruit	Change in Stem Diameter (cm) at First Pair of Leaves
Control						
Calcium						
Nitrogen						
Potassium						
Phosphorus						
Magnesium						
Sulfur						
Iron						

Name: ______________________________ Date: ______________

Results

Using the weekly data collected by your group, plot the following information using a line graph.

1. Time (number of weeks) with plant height (cm).
   You will need to remember that time 0 is your data from your initial observation. Time is on the horizontal (X) axis of the graph and plant height (cm) is on the vertical (Y) axis of your graph. Remember to divide the squares on the Y axis to cover your full range of values. For example, if your control plant began at 4 cm and grew to 25 cm, then the scale on your Y axis should be from 0 to 30. You will need to make a line for each of the seven treatments and one for the control. Use colored pencil/pen and give each line a different color. Provide a legend at the bottom of your graph (RED: Control; BLACK: Calcium; BLUE: Nitrogen, etc.).
2. Time (number of weeks) with average stem internode distance (cm).

   Follow the directions in Step 1 and plot time in weeks on the horizontal (X) axis and average stem internode distance (cm) on the vertical (Y) axis.
3. Time (number of weeks) with stem diameter at the first pair of leaves (cm).

   Follow the directions in Step 1 and plot time in weeks on the horizontal (X) axis and stem diameter at the first pair of leaves (cm) on the vertical (Y) axis.

Discussion and Conclusions

1. Describe the changes in overall appearance of the seven treatment plants over the course of the ten-week experiment.

   Calcium

   Nitrogen

   Potassium

Name: ______________________ Date: ______________

Phosphorus

Magnesium

Sulfur

Iron

2. How might these changes be related to the metabolic function of the macronutrient in plant cells?

Name: ______________________________ Date: ______________

3. Were your results different from your original predictions?

4. Were your hypotheses supported or refuted by the data your class collected?

Summarize your results in terms of implications for farming practices. How might soil pH affect your fertilizing schedule and/or use of "green" fertilizers?

Name: ______________________________ Date: ______________

Exercise 5-2: What's in a Fertilizer Bag?

(by Carroll L. Shry, Jr.)

Purpose

To acquaint students with fertilizer nutrients.

Materials

Bag of fertilizer
Bag of cottonseed meal
Bag of bone meal
Bag of blood meal
Urea
Ammonium nitrate
20-20-20 soluble fertilizer

Procedure

Fertilizer is material applied to growing media (soil) to supply the elements the plant needs for optimum growth.

Commercial fertilizers have macronutrients of three major nutrients—nitrogen, phosphorus, and potassium. These nutrients are represented by the letters N, P, K, respectively. Nitrogen supplies the element that keeps the plant color dark green during the growth period. Phosphorus encourages flower and root development and increases disease resistance. Potassium ensures starch formation, plant disease resistance, and production of chlorophyll.

The fertilizer analysis is the relationship of each macronutrient, N-P-K, to one another expressed in the form of a percentage. For example, fertilizer analysis of 10-10-10 has 10 percent nitrogen, 10 percent phosphorus, and 10 percent potassium.

1. The instructor will provide a bag of fertilizer with the label.
   a. List the plant food elements ______________________________
   b. Record the percentage of the plant food elements ______________________________
   c. Describe the physical features of the plant food ______________________________

2. Using the bag of soluble fertilizer 20-20-20:
   a. Mix a sample of the soluble fertilizer at 100 ppm (parts per million) and 200 ppm. Use the recommendations on the bag for mixing.
   b. Note the color concentration of each. ______________________________

3. How many pounds of each macronutrient of N-P-K are included in the following fertilizers?

Example: A 50 lb. bag of 10%-10%-10% fertilizer
50 lb. × .10 = 5 lb. each of N-P-K

Using this formula, determine the pounds of active ingredient in each of the following fertilizers.

Name: ______________________________ Date: ______________

Pounds of Fertilizer	Fertilizer Analysis	Pounds of Nitrogen	Pounds of Phosphorus	Pounds of Potassium
50 lb.	5-10-5			
20 lb.	23-7-7			
100 lb.	0-30-0			
40 lb.	20-20-20			
75 lb.	30-0-12			
60 lb.	12-6-6			
85 lb.	4-12-6			
35 lb.	16-8-4			
59 lb.	32-6-8			
62 lb.	12-32-8			

Answer the Following Questions

1. List two organic fertilizers. ______________________________
2. List two chemical fertilizers. ______________________________
3. What element source is supplied by urea? ______________________________
4. If urea has 45% nitrogen, how many pounds of N are in a ton? ______________________________
5. What does 20-20-20 mean? Give the fertilizer ratio. ______________________________
6. How many pounds of nitrogen are in 5 tons of leaf mold compost (5.0% N)? ______________ In 3 tons of sewage sludge (6.5% N)? ______________
7. What is the compound that creates the blue color in the solution of the 20-20-20 soluble fertilizer? ______________________________
8. Explain what nitrogen, phosphorus, and potassium do for plants. ______________________________
9. Plant food may be divided into two groups. What are they? ______________________________
10. What is the difference between major and minor elements? ______________________________

Name: ______________________________ Date: ______________

Exercise 5-3: Water Movement in Plants

(by Walter York)

Purpose

To determine water movement in plants (osmosis and diffusion).

Materials

Osmosis and diffusion kit

Procedure

1. Study the Special Notes section of the osmosis and diffusion kit instructions.
2. Fill the plastic container with water to within 2 cm from the top.
3. Test the water for presence of glucose as outlined in the osmosis and diffusion kit.
4. Tie a knot very tightly approximately 1 cm from the end of the thoroughly moistened and opened membrane tube.
5. Put 2.5 to 3 cm of glucose solution into the membrane tube.
6. Put 2.5 to 3 cm of liquid starch into the membrane tube.
7. Hold the top of the membrane tube closed by pinching it together and rinse it under running water to remove any glucose or starch on the outside. Gently squeeze the tube to mix the contents.
8. Place the filled membrane tube into the container of water, being careful not to spill the contents of the tube, and to keep the open end of the tube outside of the container.
9. After five minutes, look at the contents inside of the membrane tube and at the liquid in the container. Record your observations in Observation 1.
10. Test the water in the container for glucose, as was done in Procedure 3. Record your results in Observation 2.

Observations

1. Record the color of the liquid inside the membrane tube and of the liquid in the container.

Color of Liquid in the Membrane Tube	
Color of Liquid in the Container	

2. Record the results of the glucose test below. Circle the correct response.

   Positive

   Negative

Name: ______________________________ Date: ______________

Conclusions

1. What happened to the starch?

2. What happened to the glucose?

3. What materials diffused through the membrane?

4. How was this proven?

5. What substance did not diffuse through the membrane?

Name: ______________________________ Date: ______________

Exercise 5-4: Calculating Fertilizer and Chemical Ratios

(by Steven E. Newman)

Purpose

- To learn how to calculate fertilizer formulations in respect to soluble feed solutions used in greenhouses.
- To learn how to calibrate fertilizer and chemical ratios for injection equipment used in greenhouses.

Materials

Pen or pencil, paper, and simple calculator
20-10-20 soluble fertilizer
15-5-15 soluble fertilizer
Electrical conductivity (EC) meter
Scale
2 500-ml beakers
1,000-ml graduated cylinder
5-gallon bucket with 2-liter increments marked
Hozon™ Brass Siphon Mixer (available at most garden centers)
Dosatron® or Dosmatic® fertilizer injector (optional)
Garden hose
Water source for garden hose

Background

Commercial greenhouse growers regularly use liquid feed fertilization programs. Many growers use either a constant feed program, injecting fertilizer at each irrigation application, or a pulse feed program, injecting fertilizer on a regular periodic schedule. The program selected is determined by crop requirements, available equipment, and personal preference. The most important concern, no matter which program is used, is accuracy in calculation of fertilizer concentrations.

Crop nutrition requirements and most published fertilizer schedules use the terminology *parts per million* (ppm) to denote concentrations of specific nutrients. Fertilizer tables are provided by most fertilizer producers to convert the amount of fertilizer needed per gallon of water to generate a specific ppm concentration. With a little knowledge, a calculator, and patience, the tables should not be necessary.

Procedure 1

Many growers are familiar with the "Quick 75" method for calculating ppm. To calculate the amount of fertilizer required, divide the desired ppm by 75 and then divide by the decimal fraction of the desired nutrient (such as nitrogen, potassium, or phosphorous) contained in the fertilizer. This results in the number of ounces of fertilizer to use in 100 gallons of water.

$$\frac{\text{desired ppm} \div 75}{\text{dec. frac. of nut}} = \text{oz. of fertilizer carrier per 100 gallons of water}$$

To use this equation, assume that the fertilizer recommendation calls for 200 ppm of nitrogen from ammonium nitrate (33% N). Using the above equation, dividing 200 ppm by 75 results in 2.67, then divide this by 0.33. The answer is 8.09 ounces of ammonium nitrate, which when dissolved in 100 gallons of water will yield a solution containing 200 ppm nitrogen.

Name: ______________________________ Date: ______________

$$\frac{200 \div 75}{0.33} = 8.08 \text{ oz. of fertilizer carrier per 100 gallons of water}$$

For those who can think in metric terms, there is an easier way to calculate ppm. By definition, 1 milliliter (ml) of water weighs 1 gram (g), therefore 1 liter (1000 ml) weighs 1,000 g. Thus 1 liter of water weighs 1,000,000 milligrams (mg). This tells us that 1 ppm equals 1 mg/1,000,000 mg of water or 1 mg/liter of water. To calculate ppm in liters, simply divide the desired ppm by the nutrient fraction of the fertilizer. This results in the number of mgs of fertilizer to use in 1 liter of water.

$$\frac{\text{desired ppm}}{\text{dec. frac. of nut.}} = \text{mg of fertilizer carrier per liter of water}$$

To illustrate this equation, use the same fertilizer recommendation as before: 200 ppm N from ammonium nitrate (33% N). Substituting in the above equation, simply divide 200 by 0.33. The answer is 606 mg of ammonium nitrate, which dissolved in 1 liter of water will yield 200 ppm nitrogen.

$$\frac{200}{0.33} = 606 \text{ mg of fertilizer carrier per liter of water}$$

To increase this to irrigation volumes, multiply this result by the required volume. Multiply 606 mg by 378 liters (100 gal), which equals 229,068 mg fertilizer per 378 liters of water, or 8.08 oz. fertilizer per 100 gallons of water.

Complete the Following Exercises

1. How many ounces of fertilizer will be required to make 100 gallons of fertilizer solution at 200 ppm nitrogen (N) using 20-10-20 soluble fertilizer?

   ________ oz./100 gallons of water

2. Using your answer from question 1, how may ppm of phosphorus (P_2O_5) will be provided from the 20-10-20 fertilizer dissolved in 100 gallons of water?

   ________ ppm

3. How milligrams of fertilizer will be required to make 1 liter of fertilizer solution at 200 ppm nitrogen (N) using the 20-10-20 soluble fertilizer?

   ________ mg/liter of water

4. Using your answer from question 3, how many grams of fertilizer will need to be dissolved if you were to mix 100 liters of fertilizer solution?

   ________ g/100 liters of water

5. How many ounces of fertilizer will be required to make 200 gallons of fertilizer solution at 200 ppm nitrogen (N) using the 15-5-15 soluble fertilizer?

   ________ oz./100 gallons of water

Name: ______________________________ Date: ______________

Bonus question:
How much potassium nitrate (13-0-44) and ammonium nitrate (33-0-0) will be required to provide a concentration of 200 ppm nitrogen (N) and 200 ppm potassium (K_2O) to be dissolved in 100 gallons of water?

________ oz. potassium nitrate/100 gallons of water

________ oz. ammonium nitrate/100 gallons of water

Fertilizer Injectors/Proportioners

Background

Liquid feed fertilizer programs used in commercial greenhouse crop production allow a grower to manipulate plant nutrition conveniently. This is done by using different fertilizers containing equal ratios of nutrients and adjustment of specific fertilizer elements within a fertilization program. These ratios are calculated using the formulas discussed previously.

Fertilizer injectors or proportioners available with several dilution ratios are used in greenhouses to deliver large volumes of fertilizer in irrigation water by injecting a small amount of concentrated fertilizer solution into the irrigation water. The most commonly used injection ratio is 1:100. This means that 1 unit of concentrated fertilizer stock solution is combined with 99 units of water. Thus, a fertilizer solution with a 1:100 dilution is added to the crop. The volume of the unit may be a pint, quart, or gallon or liter. The unit may be any volume as long as both the fertilizer and the water are measured in the same unit.

The Hozon™ Brass Siphon Mixer, which is available at most garden centers, is an inexpensive venture-type injector used by small growers for small areas. Its dilution ratio is typically 1:16.

Procedure 2

1. Connect the Hozon™ Brass Siphon Mixer to the water faucet and then connect the garden hose to the Hozon™ Brass Siphon Mixer. Place the siphon tube into a container of water and run the system until water is drawn into the siphon tube; turn the water faucet off.
2. Fill a 500-ml graduated cylinder to the 500-ml mark with water.
3. Place the siphon tube into the 500-ml graduated cylinder and the hose end into a 5-gallon bucket and turn on the water.
4. As accurately as you can, collect 6 liters from the hose in the 5-gallon bucket.
5. Determine how many milliliters of water were removed from the graduated cylinder.

   ________ ml of water removed from graduated cylinder

   ________ liters of water collected in 5-gallon bucket
6. What is the ratio of water removed from the cylinder to the volume of water collected in the bucket?

   ________:________
7. How does this value differ from the expected 1:16, and if different, why?

   ____

   ____

   ____

Name: ______________________________ Date: ______________

8. How many ounces of 20-10-20 fertilizer at 200 ppm nitrogen (N) will be required to mix in 1 gallon of fertilizer concentrate delivered using a 1:16 injector?

_________ oz./gallon of fertilizer concentrate

HINT: One gallon of concentrate using a 1:100 injector has the capacity to provide 100 gallons of diluted fertilizer and irrigation water.

9. How milligrams of fertilizer will be required to make 400 ml of fertilizer concentrate delivered using a 1:16 siphon mixer that will yield a fertilizer solution at 200 ppm nitrogen (N) using the 20-10-20 soluble fertilizer?

_________ mg/400 ml fertilizer concentrate

10. How milligrams of fertilizer will be required to make 400 ml of fertilizer concentrate delivered using a 1:16 siphon mixer that will yield a fertilizer solution at 200 ppm nitrogen (N) using the 15-50-15 soluble fertilizer?

_________ mg/400 ml fertilizer concentrate

11. Determine the EC of the clean faucet water.

_________ dS/m

12. Weigh out the amount of 20-10-20 from question 9 with 400 ml water in a 500-ml beaker. Place the siphon tube into the 500-ml beaker and turn on the water on. Once the flow of water has stabilized and dilute fertilizer is coming from the end of the hose, collect a sample of the water and determine its electrical conductivity with the EC meter. What is the EC of the dilute fertilizer?

_________ dS/m

13. Repeat question 12 using the amount of 15-5-15 fertilizer from question 10. What is the EC of the dilute fertilizer?

_________ dS/m

14. How do your results differ from the results of questions 9 and 10? Why?

____

____

____

SECTION

6

Plant Propagation

Notes

Name: ______________________________ Date: ______________

Exercise 6-1: Plant Reproduction

(by Catherine A. Teare Ketter)

Purpose

To study the reproductive cycles of plants and to understand the various mechanisms of pollination and seed production. Specifically, you should

- Distinguish between vascular and nonvascular plants and give an example of each type of plant.
- Recognize the male and female cones of a coniferous gymnosperm.
- Draw and label the stages in the life cycle of a pine.
- Define angiosperm, and describe the sexual reproductive cycle of angiosperm.
- Identify the male and female gametes of an angiosperm.
- Relate the shape of the flower to the method of pollination.
- Describe the various mechanisms of pollination found in angiosperms and describe the economic/agricultural implications.
- Relate seed shape to seed dispersal.
- List some mechanisms for the asexual propagation of flowering plants.

Materials

Procedure 2
Preserved staminate cone
Preserved ovulate cone
Desiccated ovulate cone
Alcohol/glycerol solution
Oil immersion microscope

Procedure 3
Monocot and dicot flowers
Razor blade
Dissecting microscope

Procedure 4
Pollination gel
Distilled water
Toothpicks
Microscope
Slides and coverslips

Procedure 5
Immature okra pods

Procedure 6
Display of asexually propagated plants

Name: ______________________________ Date: ______________

Procedure 1

Plants play an extremely important role in the balance of nature. Plants modify both local and global climate patterns by adding water vapor into the atmosphere through a process called transpiration. Plants also release oxygen (O_2) into the atmosphere. The oxygen is a by-product of photosynthesis, and oxygen is required by all animals to sustain life. Additionally, plants provide food for most animals, some bacteria, and some fungi. Humans have cultivated plants for several centuries to produce food. Studying the reproductive cycles of plants allows the commercial and home producer to maximize crop yield by understanding pollination mechanisms, fruit set, and seed production.

Members of the plant kingdom can be divided into two groups, vascular and nonvascular plants. Vascular plants have specialized conductive tissues which transport water and nutrients through the plant. Ferns, pine trees, tomato, cotton, and tobacco are all examples of vascular plants. Ferns have vascular tissue only in the stems and leaves. The roots lack vascular tissue and are called rhizoids. Pine trees have "true" roots, stems, and leaves, all possessing vascular tissue. In contrast, the simplest members of the plant kingdom lack vascular tissue. The bryophytes (mosses and liverworts) lack vascular tissue. Water and nutrients are transported between cells by the processes of osmosis and diffusion.

Hypothesis

Write a hypothesis that relates the presence or absence of vascular tissue to plant size. Give examples of plants that fit your hypothesis.

Name: ______________________________ Date: ______________

Procedure 2—Conifers

Members of the Coniferophyta are abundant. Some 550 species have been described. Conifers are gymnosperms, plants which produce seeds lacking the protective integuments of angiosperm seeds. The Coniferophyta are found in regions of the world in which length of the growing season is shortened due to latitude or longitude. Most members of the Coniferophyta are evergreen, but a few deciduous members can be found in semitropical and tropical areas of the world. Permanent leaves allow conifers to take advantage of any sunlight available for photosynthetic activity. The shape of the needles and growth habits of the northern and southern temperate latitude trees allow conifers to withstand large snow loads without losing branches. Conifers are economically important; many of the trees which supply lumber and pulp are conifers—pine, hemlock, cypress, fir, and spruce.

1. Male gametophyte—*Pinus*

The male gametophyte is produced in a staminate cone borne at the terminal portion of branches (see Figure 6-1A). Meiosis occurs to produce microspores within each segment (scale) of the male cone. The microspores of different conifer species vary in size and shape, modified to enhance pollination in different habitats. Virginia pine (*Pinus virginiana*) has pollen grains which look like "Mickey Mouse" hats (see Figure 6-1B). Texas pine (*Pinus taeda*) has pollen grains with air bladders.

Hypothesis

Write a hypothesis that relates the shape of conifer pollen to the method of pollen transport. (HINT: Think about the shape.)

Why do you think that conifer pollination occurs in the spring instead of the fall or winter? (HINT: Think about the typical climate of each season.)

Name: ______________________________ Date: ______________

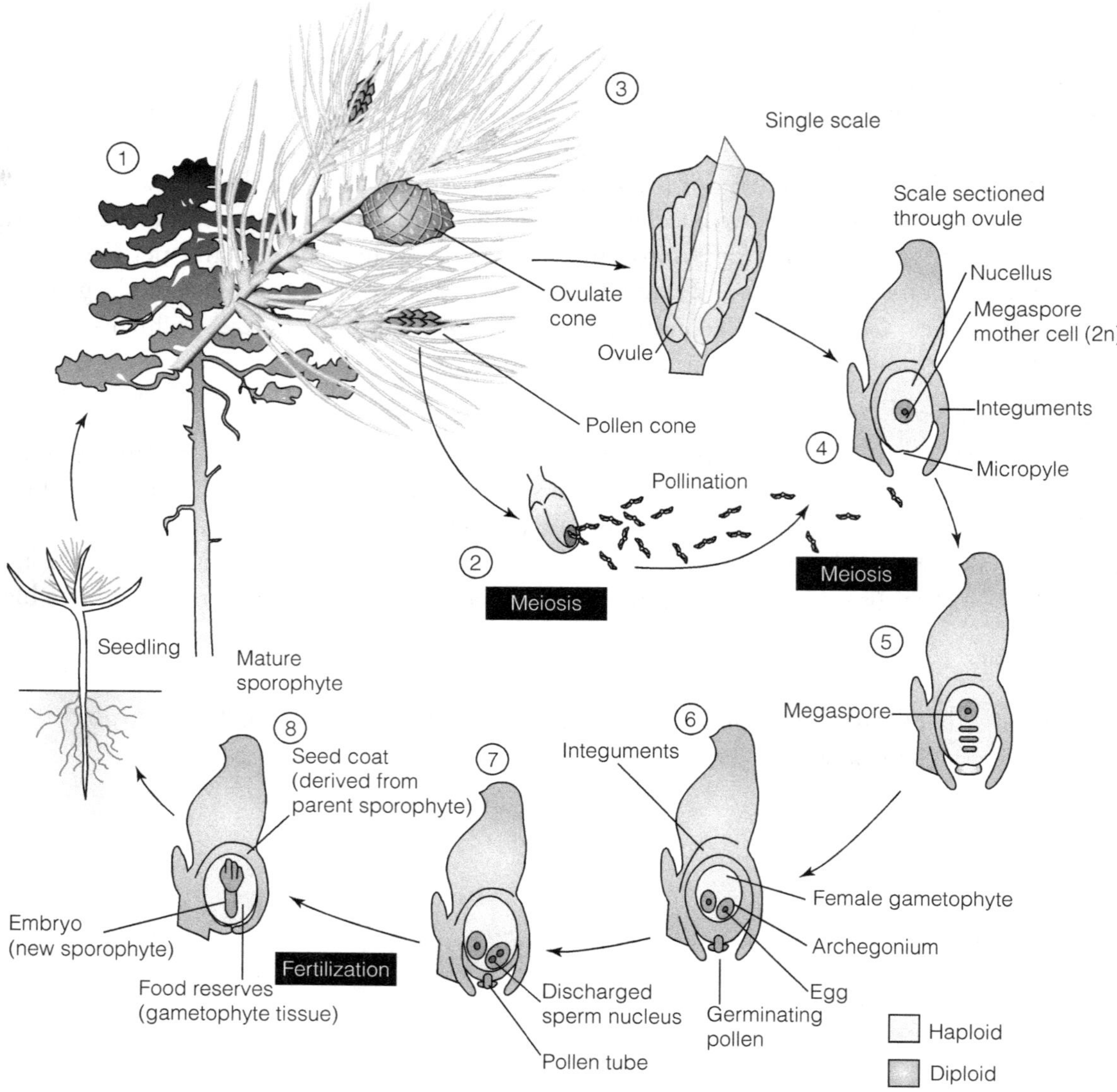

Figure 6-1A Life cycle of a pine.

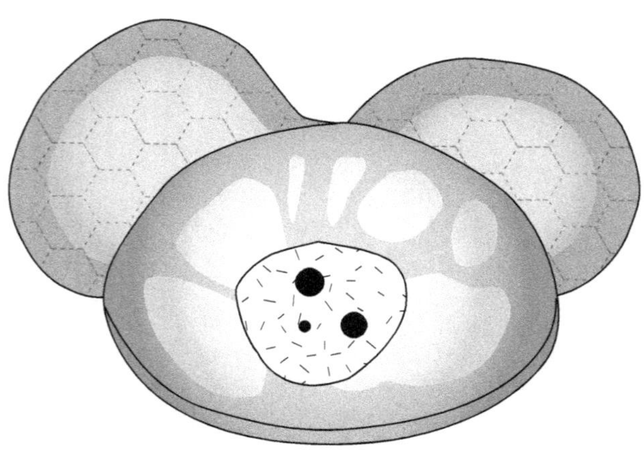

Figure 6-1B Virginiana pine pollen.

Name: ____ Date: ____________________

Obtain a preserved staminate cone and remove a scale from the side of the male cone. Mount the scale in an alcohol/glycerol solution on a slide. Crush the scale and apply a coverslip. Examine the slide using the 10×, 40×, and oil immersion objectives. (NOTE: Use the oil immersion lens only if the preparation is thin enough to accommodate the lens.) Draw and label what you observe in the space provided.

What is the purpose of the "wings" on the *Pinus* pollen? How is the shape of the pollen grain related to reproductive success?

2. Female gametophyte—*Pinus*

The female gametophyte is produced in the ovulate cone at the apex of lateral branches. The female ovulate cones are produced on the same tree as the male staminate cones; *Pinus* is monoecious. Large scales are arranged in whorls around a central axis. The ovule is at the base of the scale (see Figure 6-1A).

Obtain a preserved ovulate cone and remove one of the small young ovulate scales at the top of the cone. Place the scale in a depression slide in a solution of water and place a coverslip on the slide. Using the 10× objective, draw and label what you see in the following space.

Name: ______________________________ Date: ______________

What is the purpose of the scales and bracts of the female cone?

Why are the staminate cones borne on the ends of terminal branches and the ovulate cones on lateral branches?

3. Seeds and seed dispersal—*Pinus*

The seeds of the pine are released from the female cone after one to two seasons of development, depending upon the species. Seeds of the pine are released as the cone desiccates (see Figures 6-1A and 6-1C).

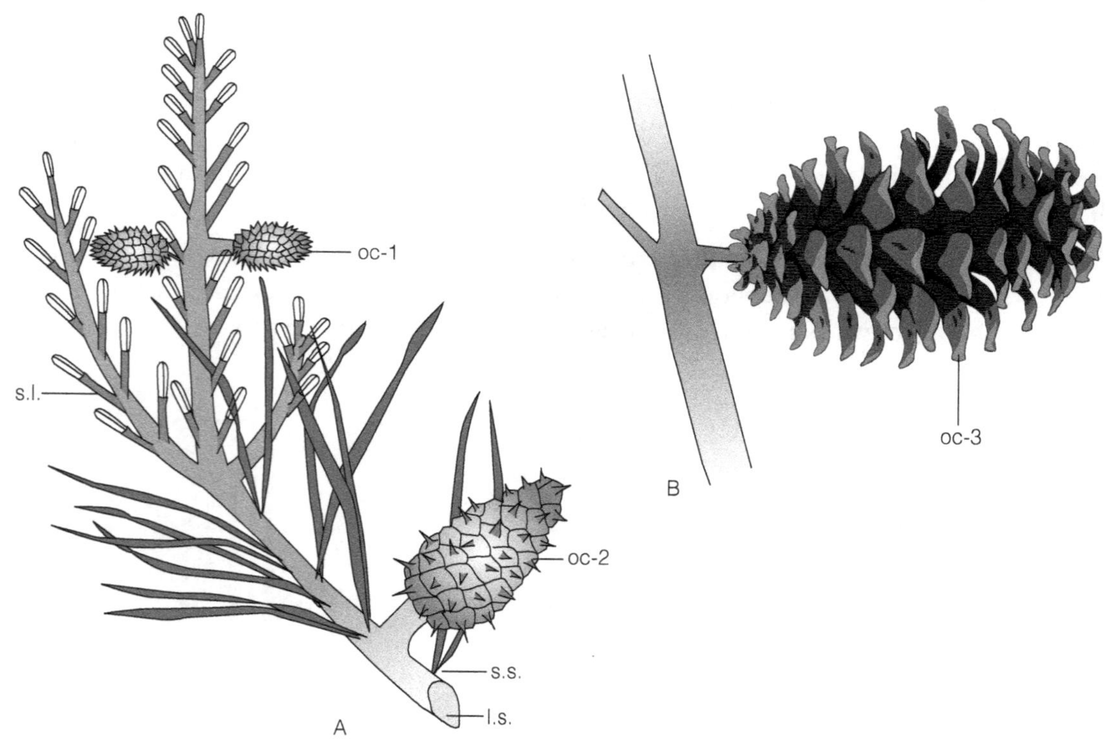

Pinus virginiana Miller. (A) Long shoot with ovulate cone in spring, just after pollination. (B) Ovulate cone shedding seed. l.s., long shoot; oc-1, ovulate cone of the season, just after pollination; oc-2, ovulate cone pollinated 1 year earlier; oc-3, ovulate cone pollinated 2 years earlier; s.l., scale leaf; s.s., spur shoot. x1.

Figure 6-1C One, two, and three season ovulate cones of the Virginia pine.

Name: ______________________________ Date: ______________

Some seeds of western conifer species are released only after a fire. What advantage would fire give a seedling?

What role could controlled burning play in the management of commercial conifer stands?

Observe the various pine cones on display. Note the diversity of shape and size among cones of various conifers. To observe the effects of water loss and hydration on a mature ovulate cone, a desiccated cone is placed in a pan of water at the beginning of the lab period. Record initial observations on the number and the size of openings between the scales (use a metric ruler to make the measurements).

Observation of Pine Cone

	Initial (dry)	**Final (wet)**
Number of open scales	________	________
Average distance between scales	________	________

After an hour, how has the shape of the cone changed?

Name: ______________________________ Date: ______________

Procedure 3—Flower Structure

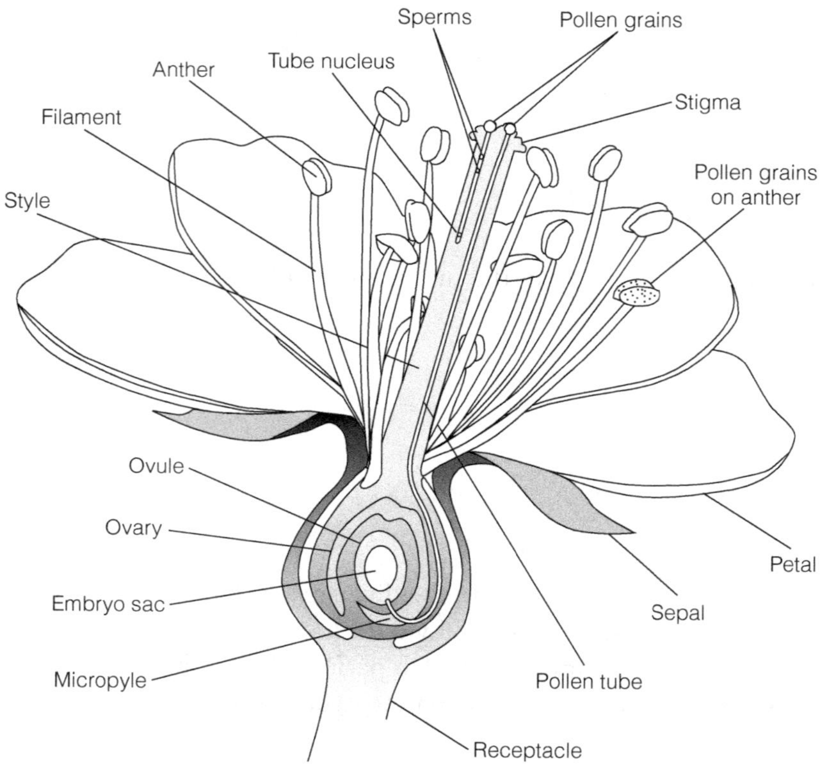

Figure 6-1D Parts of a flower.

1. Working in groups of two, obtain a fresh monocot flower (a gladiolus) and a dicot flower (a petunia or tobacco) for dissection.
2. First, examine the monocot flower. Note that the two leaflike structures at the base of this flower are not sepals but are actually modified leaves called bracts. The sepals are the three outer petal-like appendages which are called tepals when they look like petals, as in this flower.
3. Slice lengthwise through the tepals and ovary with a razor blade, so that the reproductive structures can be viewed in a longitudinal section. Count the floral parts and draw the monocot flower, identifying the parts using Figure 6-1D.
4. Make a cross section (across the long axis) through the ovary of the gladiolus and examine it under the dissecting microscope. Draw the section viewed and label the ovules.

Name: ______________________________ Date: ______________

Cross section of monocot flower ovary

5. Follow the same procedure to dissect the dicot flower.

Cross section of dicot flower ovary

Flowers that are bright red or orange with tubular shapes are frequently pollinated by hummingbirds. Many plants which are pollinated by bees are bright yellow, blue, or purple and have strong, sweet scents. If these flowers are examined under ultraviolet light, markings which guide the bee to the anthers are evident. Some flowers have evolved unique odors to attract pollinators. A cactus which is pollinated by carrion beetles produces a flower which smells like rotting flesh. Many flowers with obnoxious odors are pollinated by beetles and flies that live in rotting flesh or dung. Some flowers are pollinated at night by nocturnal bats and moths. The yucca plant is pollinated by a yucca moth.

Observe the flowers on demonstration and suggest possible pollinator.

daylily	______________	red salvia	______________
buttercup	______________	night blooming cactus	______________
orchid	______________	rose	______________

Your teacher may show you some kodachrome slides of other flowers, some of which may be of commercial importance. Write the names of the plants and their pollinators in the space provided.

Procedure 4—Pollen Tube

1. Apply a film of pollination gel to the center of a slide. Close gel jar.
2. Hold slide under yellow anthers of *Tradescantia* flowers. Please *do not remove* the entire anther; follow step 3.
3. Tap anthers twice, gently, to release pollen. Place slide on black surface to see pollen.
4. Add 1 drop of water. Mix pollen, water, and gel, using a toothpick. Mixture is lumpy.
5. Drop coverslip onto slide. Don't squash.

   This may be the only slide you'll ever make in which it is advantageous to have air bubbles. Air bubbles provide oxygen during germination.

Name: ______________________________ Date: ______________

6. Use moderate light level on microscope. Do not overheat slide (microscope lights are hot).
7. View using the 10× objective until you see pollen grains ("footballs").
8. Switch to the 40× objective when germination begins, and observe.
9. Re-examine at 15-minute intervals to observe a germinating pollen tube. Turn off microscope light between viewings.
10. Add distilled water to the edge of the coverslip as needed to rehydrate the preparation.
11. Look for the nucleus in the tube when the pollen tube has germinated. Draw the pollen grain, showing the surface texture and the tube.

Procedure 5—Fruit Structure

1. Obtain an immature okra pod (edible at this stage). Make a cross section through the fruit and draw the developing seeds in the pod.
   How does the okra pod compare with the gladiolus ovary?

2. Draw a mature okra fruit and indicate how the seeds are released from the pod.

Fruits are often referred to as vegetables, even though this is not botanically correct. The term vegetable refers to a vegetative (non-reproductive) part of the plant. Identify the following items as a vegetable (as used botanically) or as a fruit.

green beans	______________	lettuce	______________
carrots	______________	squash	______________
turnips	______________	peas	______________
spinach	______________	watermelon	______________
okra	______________	eggplant	______________
white potatoes	______________	beets	______________

How are the following plants adapted for seed dispersal?

tomato	______________	coconut	______________
maple	______________	milkweed	______________
cocklebur	______________	cherry	______________

Name: ______________________________ Date: ______________

Hypothesis

Write a hypothesis that relates seed shape and size to the type of dispersal mechanism employed by the plant. Support your hypothesis with evidence from your "Fruit Structure" observations.

Procedure 6—Asexual Propagation

Sexual reproduction in flowering plants produces seeds. The plants which develop when the seeds germinate are genetically different from the parent plant(s). Sexual reproduction increases the genetic variation within a population of plants. Plants are specifically bred for certain desirable characteristics: sweetness in corn, nematode resistance in cotton, fusarium wilt resistance in tomatoes, and mosaic virus resistance in tobacco are examples of highly desirable traits. In such instances, it might be preferable to produce plants that are exact genetic copies of the parent and thus contain all of the desirable traits. Asexual propagation produces new plants which are identical to the parent plant.

Look at the asexually propagated plants on display. Indicate the type of propagation used next to the name of the plant.

garlic	____________	Easter lily	____________
zoysia (grass)	____________	lily of the valley	____________
crocus	____________	chives	____________
tobacco	____________	orange	____________
costal bermuda	____________	bahia	____________
bugleweed (ground cover)	____________	jade plant	____________

Name: ______________________________ Date: ______________

Table 6-1A

Types of Asexual Propagation		
Mode of Reproduction	**Species**	**Description**
PROPAGATION TECHNIQUE		
Parthenogenesis	rose, orange	Embryo develops from unfertilized haploid egg or from tissue surrounding the embryo sac
Tissue culture	carrot, corn, wheat, rice, orchid	New plant arises from undifferentiated tissue derived from the parent plant—laboratory technique
Vegetative	African violet, hosta	New plant develops from tissue which is separated from parent plant
REPRODUCTION BY MODIFIED STEMS		
Bulb	onion, tulip	Axillary bud on underground stem gives rise to new bulb
Corn	gladiolus	Axillary bud on short, thick underground stem produces new plant
Rhizome	running grasses	New plants develop at nodes on underground stems
Runner	strawberries	New plants develop from nodes on horizontal surface roots
Tuber	potato	Enlarged tips of underground rhizomes produce new plants

Under what conditions do you think that increased genetic diversity of a crop plant would be desirable?

What is the relationship between monoculture growing practices and the need for the development of new and improved strains of crops/plants?

Name: ______________________________ Date: ______________

Exercise 6-2: Requirements for Seed Germination

(by Walter York)

Purpose

To determine how differences in environmental conditions affect seed germination.

Materials

Water
Aluminum foil
Beaker
Hot plate
Freezer
Seed-Gro kit

Procedure

Moisture Experiment

1. Place a piece of dry aluminum foil in a dry place.
2. Place five seeds on the foil or paper. *Do not water.*
3. Place one of the foam germination pads in the petri dish and water as directed by your instructor.
4. Place three seeds of the same kind as you placed in the foil on the foam pad and add a few drops of water directly on the seeds to hasten germination.
5. Replace the cover on the germination chamber to reduce moisture loss and keep out unwanted materials.
6. Observe the seeds for a seven-day period. Record your observations on the chart in Observation 1.

Air (Oxygen) Experiment

1. Fill a beaker approximately half full with water.
2. Place five seeds in the beaker.
3. Place one of the foam germination pads in the petri dish and water as directed by your instructor.
4. Place three seeds of the same kind as you placed in the beaker on the foam pad and add a few drops of water directly on the seeds to hasten germination.
5. Replace the cover on the germination chamber to reduce moisture loss and keep out unwanted materials.
6. Observe the seeds for a seven-day period. Record your observations on the chart in Observation 2.

Temperature Experiment

1. Prepare two packets each containing five seeds wrapped in aluminum foil.
2. Place one packet in the freezer for one week.
3. Place the other packet in boiling water for ten minutes.
4. Prepare three germination chambers by placing the foam germination pads in the petri dishes and watering as by your instructor.

Name: ______________________________ Date: ______________

5. Label the germination chambers "hot," "cold," and "normal."
6. Plant the seeds boiled in water in the chamber labeled "hot," the freezer-conditioned seeds in the chamber labeled "cold," and the five untreated seeds in the chamber labeled "normal."
7. Observe the seeds for a seven-day period.
8. Record your observations on the chart in Observation 3.

Light Experiment

1. Prepare two germination chambers by placing the foam germination pads in the petri dishes and watering as directed by your instructor.
2. Plant three seeds of the same kind in each chamber.
3. Wrap one chamber in aluminum foil, making sure no light can get into the chamber.
4. Observe the seeds every day for one week. Try to let no light (or as little as possible) get to the seeds in the aluminum-foil-wrapped chamber when you observe them.
5. Record your findings on the chart in Observation 4.

Observations

1. Record your moisture observations in the following chart:

	Day 1	Day 2	Day 3	Day 4	Day 5	Day 6	Day 7
Germination Chamber Seeds							
Foil-Wrapped Seeds							

2. Record the results of the air experiment in the following chart:

	Day 1	Day 2	Day 3	Day 4	Day 5	Day 6	Day 7
Germination Chamber Seeds							
Seeds in Beaker of Water							

Name: ______________________ Date: ______________

3. Record the results of the temperature experiment in the following chart:

	Day 1	Day 2	Day 3	Day 4	Day 5	Day 6	Day 7
Normal Seeds							
Hot Seeds							
Cold Seeds							

4. Record the results of the light experiment in the following chart:

	Day 1	Day 2	Day 3	Day 4	Day 5	Day 6	Day 7
Germination Chamber Seeds							
Foil-Wrapped Seeds							

Conclusions

1. What part does moisture play in seed germination?

2. What happened to the seeds that were placed in the beaker of water? What caused this?

Name: ______________________________ Date: ______________

3. What effect did heat have on the germination of the seeds?

4. What effect did cold temperature have on the germination of the seeds?

5. How did the absence of light affect the germination of the seeds?

6. What conditions are necessary for proper seed germination?

SECTION

7

Container Grown Crops

Notes

Name: ______________________________ Date: ______________

Exercise 7-1: Planting Container Plants with the Butterfly Technique

(by Carroll L. Shry, Jr.)

Purpose

To learn how to plant container plants using the butterfly technique to ensure the proper root growth development in landscape plants.

Materials

Nursery spade
Root-bound container plant
Knife

CAUTION
Use care when handling a sharp knife.

Procedure

1. Remove the plant from the container. The plant may be difficult to remove because of root growth.
2. Use a knife or nursery spade to cut through the bottom half of the root ball.
3. Butterfly the root ball to bring the root system closer to the surface. (Your instructor will demonstrate the butterfly techniques.)
4. Plant shrubs 1 to 2 inches above the existing grade.

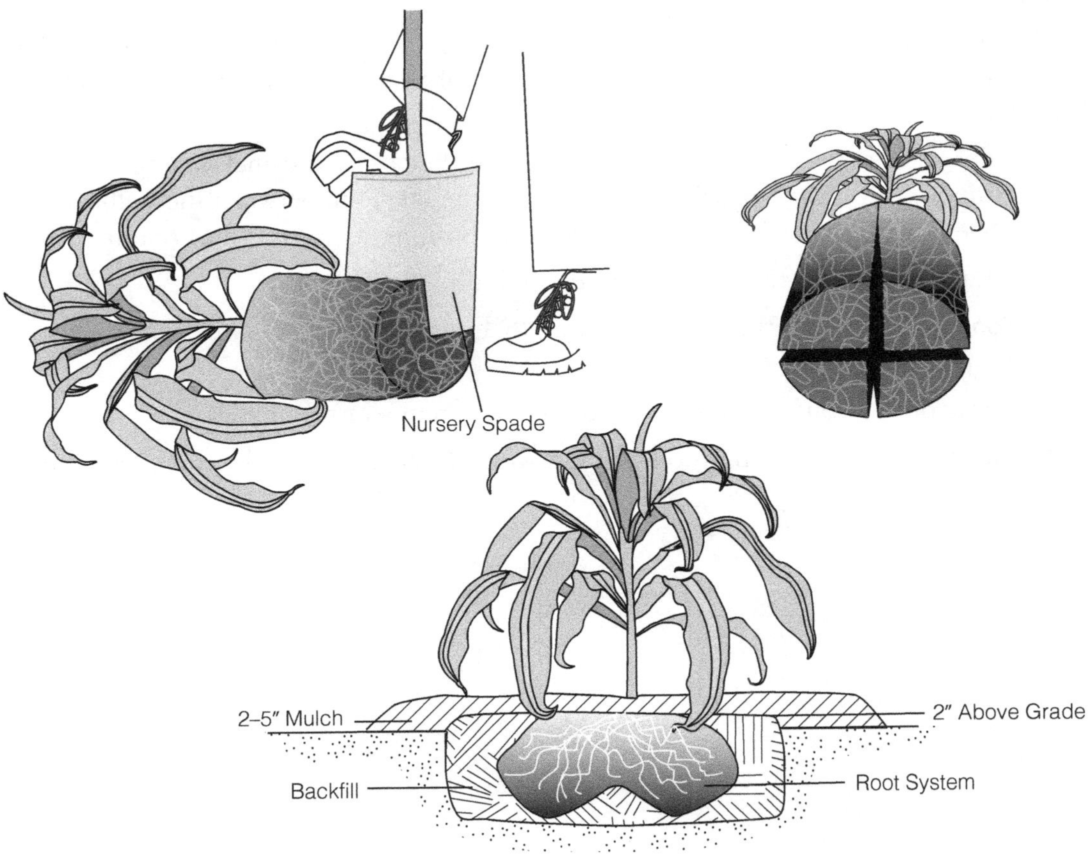

Name: ______________________________ Date: ______________

Exercise 7-2: Forcing Bulbs for Spring

(by Carroll L. Shry, Jr.)

Purpose

To force hyacinths and narcissi bulbs to bloom for Valentine's Day.

Materials

Precooled bulbs from a bulb wholesaler of hyacinths and narcissi
Bulb pans (shallow pans)
Planting media (loam soil or greenhouse media)
Plant labels
Rooting rooms to regulate temperature and moisture (48°F)
Greenhouse for forcing into bloom

Procedure

1. Select the hyacinths and narcissi that will be used for forcing.
2. Buy precooled bulbs from the wholesaler.
3. Plant bulbs mid-October into bulb pans.
4. Fill the bulb pan to the level that will allow the tops of the bulb to be exposed when planting has been completed.
5. Clean the outer tunic layer (bulb skin). Be careful not to damage the basal plate where the roots develop during the cooling treatment. (Note: the skin of hyacinths may cause some irritation to sensitive skin. Wetting the hands will also help.)
6. Place the bulbs so they are not touching each other in the pot. In a 6-inch pan, plant six narcissi bulbs. In a 6-inch pan, plant 4 hyacinth bulbs. Work the media in between the bulbs so that only ¼ of the bulb is exposed. This will keep the bulbs from pushing up during the cooling process. Water.
7. Write a plant pot label with the date the variety of bulb was planted.
8. Place in a rooting room with the temperature at 48°F for the first four weeks and then lowered to 41°F for the next ten weeks.
9. Check regularly on the soil moisture and rooting room temperature.
10. Move the bulbs out of the rooting room the last week of January for Valentine's Day flowering.

Name: ______________________________ Date: ______________

Exercise 7-3: Production of Garden Mums

(by Carroll L. Shry, Jr.)

Purpose

To propagate garden mums.

Materials

Order garden mums from a wholesale supplier
Media mix
20-10-20 soluble fertilizer
8″ mum pot (plastic or fiber)
Mum plant labels
Slow-release fertilizer
Weed barrier fiber

Garden mums give a great late summer and fall assortment of colors. Their natural season is in the fall because they are short-day plants. Today there is a variety of colors to choose from for the landscape. New flower forms and improved plant quality have made the garden mum very popular with perennial gardeners.

This crop will give students an opportunity to grow their own plants at home through the supervised agricultural experience program of the FFA.

Procedure

1. Select a plot in full sun with good drainage and cover the area with weed barrier fabric to control any weed growth.
2. Select 8″ mum pots and fill with the media mix (for example, Pro-Mix, Metro-Mix).
3. Plant one cutting in the center of the pot. (Fall crops should be started in June.)
4. Water and fertilize the mums after transplanting. Use 300 ppm of 20-10-20. Continue to feed weekly.
5. Watering is important—do not allow them to dry out. The plants will wilt, form premature buds, and reduce branching.
6. Methods of watering: a garden hose with a water breaker; drip tube system to reduce labor and conserve water; overhead watering with a sprinkler.
7. To supplement the liquid fertilizer program, use a slow-release fertilizer in the pot.
8. After two weeks pinch the mums to induce branching (breaks) and again by July 15 in the North and the end of July in the South.
9. Spacing is important for a 8″ pot. The spacing should be 14″ to 16″ on center. Proper spacing promotes air movement around the plant and optimum growth.
10. Use IPM to control diseases and insects.

SECTION

8

Cut Flower Crops

Notes

Name: ________________________ Date: ____________

Exercise 8-1: Conditioning and Storing Cut Flowers and Greens

(by Charles Griner)

Purpose

To receive a shipment of flowers and treat them in a manner that extends the keeping quality of the flowers.

Materials

Assorted flowers as they are received from wholesaler
Flower containers
Floral preservative
Knife

Procedure

Flowers and foliage delivered by a wholesale florist are shipped in large cardboard boxes and wrapped in paper for protection (Figure 8-1A). Usually a small bag of ice is also included in the box to help prevent heat buildup. However, the flowers are shipped dry and may have been in the delivery truck for many hours. The following discussion will examine the steps in conditioning a shipment of flowers.

1. *Unpack the flowers.* As soon as flowers arrive, unpack them and loosen the paper or plastic sleeves in which they have been wrapped. As flowers mature they expand. If the sleeves are not loosened, many of the flowers will be crushed as the blooms open. Do not loosen the sleeves on roses because the customers desire their flowers in the bud stage.

Figure 8-1A A flower shop employee unpacks a shipment of flowers.
(Courtesy of Steven E. Newman)

Name: ______________________________ Date: ______________

Check flowers for signs of disease, damage, or wilting. Remember, diseased and damaged flowers give off an increased amount of ethylene gas, which accelerates flower deterioration. If the number of diseased and damaged flowers is minor, remove them from the bunch before storage. If the damage is excessive, it should be reported to the supplier.

2. *Recut the stems.* The stems of all flowers should be recut before being placed in storage. This is because of the blockage that occurs in the end of the stem.

Stems should be cut with a knife rather than shears. Even sharp shears can pinch the xylem tubes causing a partial blockage. Make the second cut one to two inches above the stem base. Cut the stems on a slant. This will not help the stem absorb more water. Rather, an angle cut prevents the stems from sealing to the bottom of the container as they would with a straight cut.

Cutting the stems under warm water is beneficial (Figure 8-1B). When cut in the air, the stem may suck up a small amount of air, forming a bubble at the base, which can create a blockage. Use warm water because it contains less air than cold does. After the stems have been cut, they can be moved to storage containers. A clinging drop of water on the end of the stem will protect the new cut. Stems that are difficult to cut under water by hand may be cut using an underwater cutter (Figure 8-1C). This device cuts fresh flowers underwater by the bunch. It has a blade that sits in a tub and automatically cuts all stems on an angle.

The stems of some flowers may require special treatment. The flowers of woody stems should be cut like other flowers, but instead of cutting an inch off the bottom, move up the stem to the part that is more succulent. Succulent stems are softer and juicier and therefore better conductors of water.

Stems of plants such as poinsettias have a milky sap that bleeds after being cut. This sap reseals the cut surface and prevents the stem from absorbing water. To prevent this, burn the end of the stem over a small flame until the stem is blackened (Figure 8-1D). Another method is to place the end of the freshly cut stems in a container of boiling water for 10 to 30 seconds. Take care that the heat does not damage the foliage or flowers. This procedure causes the sap to clot and allows the stem to absorb water.

Figure 8-1B Recut stems with a sharp knife under warm water. *(Courtesy Steven E. Newman)*

Name: ______________________________ Date: ______________

Figure 8-1C An underwater stem cutter. *(Courtesy Steven E. Newman)*

3. *Remove lower foliage.* Remove all foliage from the stems that would be underwater in the storage containers. Foliage left under the water will decay, fouling the water and leading to bacterial growth. This results in clogged stems, which block water absorption. Rotting foliage also releases ethylene gas, which hastens flower deterioration.

   When removing foliage or thorns, use a heavy glove or rag. Pull the glove or rag quickly down the stem (Figure 8-1E). This does a faster, neater job than a knife, which can injure the bark of the flower. If roses are being treated, remove the outside petals of the rose also.

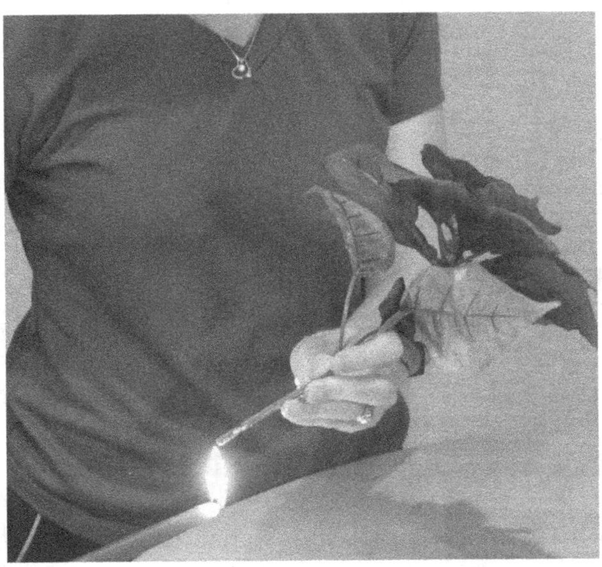

Figure 8-1D Burn the end of the stems containing a milky sap over a small flame until the stem is blackened. *(Courtesy Steven E. Newman)*

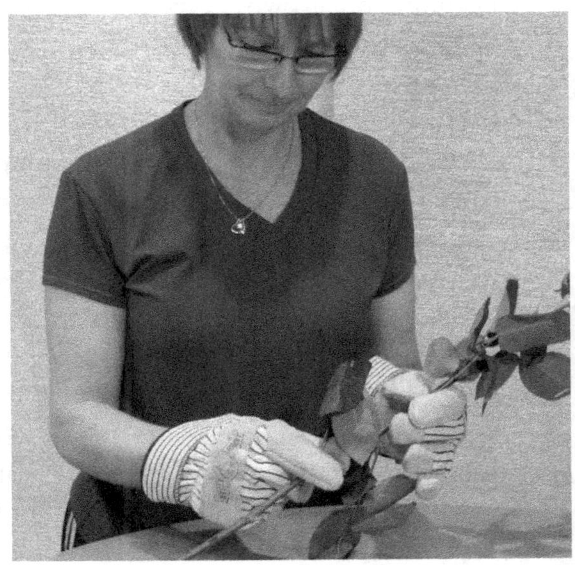

Figure 8-1E Use a heavy glove or rag to remove foliage or thorns. *(Courtesy Steven E. Newman)*

Name: ______________________________ Date: ______________

4. *Clean containers and cooler.* Once flowers have been cut and cleaned, they are ready for storage. Use containers that have been scrubbed with a hot detergent solution, disinfected with bleach, and rinsed thoroughly. A 10 percent solution of bleach can be used to disinfect the containers.

   Floralife, Inc. makes a product, Floralife Formula D.C.D., that can be used to clean and disinfect in one step. D.C.D. stands for "disinfects, cleans, and deodorizes." This material kills many bacteria, fungi, and algae that produce ethylene gas and also block flower stems. In addition, the detergents clean and deodorize container surfaces as they are disinfected.

   Regardless of the product used, sanitation is the key. Diligently scrub buckets, shelves, floors and sinks, and remove all sources of ethylene gas. These vital steps will prolong the life of your fresh flowers.

   Non-metallic containers should be used. Metal containers can decrease preservative effectiveness. To prevent flower injury, select a container that is short enough that the flowers will not come in contact with the sides of it.

5. *Use a preservative.* Place a warm preservative solution in the container prior to adding flowers. The temperature of the solution should be between 100 and 110°F. This temperature should feel very warm to your hands but not burn. Warm water should be used because it contains less air and will travel up the stem more easily.

   Preservatives added to flower water have been found to be very effective in extending the life of cut flowers. They do this in three ways.

   First, they provide a food source. As long as the flower is alive, it continues to convert stored food reserves into energy. This process is called respiration. When the stored food in the cut flower is used, the flower dies. The preservatives contain sugar that the flower uses to manufacture food to replace that lost through respiration.

   Second, the preservative contains an acidifier that lowers the pH of the water. Water moves through a flower's vascular system best at a pH of 3.5. The acidifier also helps to reduce bacterial action.

   Finally, preservatives also contain a bactericide that kills bacteria. Bacteria in the water clog the stems, restricting water uptake. They also cause the water to sour and smell bad. Additionally, controlling the growth of bacteria helps to reduce the production of ethylene gas.

   A number of commercially prepared solutions are available. These can be purchased in liquid or powder form. The directions for mixing the preservatives should be followed exactly since too much preservative can burn the flower and too little will not be enough to keep the flowers fresh. It does not matter how deep the water and preservative solution is because flowers suck up water like a straw. An increase in depth cannot move more water up the stem.

   If commercial preparations are not available, a preservative can be made at home. It is effective but requires more work and is expensive. Mix 50 percent warm water and 50 percent Sprite, 7-UP, or similar drink containing citric acid. Add $\frac{1}{12}$ teaspoons of household bleach to each quart of solution. The drink contains sugar as a source of energy, and the citric acid in the drink is an acidifier. The bleach will keep bacteria from forming in the water.

6. *Allow flowers to absorb water.* All flowers with the exception of roses should remain in the warm preservative solution outside the refrigerator for one to two hours. Roses should be stored in the cooler immediately. This will allow them to absorb a maximum amount of water. At the end of this time the flowers should feel turgid (full of water). Flowers that are shipped in bud such as gladioli, lilies, and carnations could sit at room temperature overnight to open up and reach optimum condition.

Name: ______________________________ Date: ______________

7. *Refrigerate.* The floral cooler becomes the storage place for flowers until they are passed on to the consumer (Figure 8-1F). Even if flowers are to be used quickly, they should first be cooled so that they become crisp and hard.

If the flowers become wet while being prepared, allow them to dry before being placed in the cooler. Wet flowers encourage the growth of botrytis.

The floral cooler provides a low temperature and high humidity. This combination helps to extend the life of the flowers. Low temperature slows down the activity of the flowers. The respiration and transpiration rates are lowered, as well as bacterial growth and the effects of ethylene gas. The high humidity in the cooler also slows down transpiration. This helps to prevent the flowers from drying out. Flowers last longer at 90 percent humidity.

The temperature at which the cooler should be set depends upon the kind of flowers being stored. The ideal average temperature to keep most commonly used flowers, such as roses and carnations, is 36 to 40°F (2 to 5°C).

Tropical flowers, such as anthuriums and orchids, can be damaged at temperatures below 40°F (5°C). These flowers should be stored in a separate cooler at 45 to 50°F (7 to 10°C). If only one cooler is available, the tropicals can be stored at room temperature. Select a cool area in the shop that is free of drafts.

Store flowers in the cooler in a manner that allows for good air circulation around the flowers. This will help to control diseases in the cooler. Clean the cooler regularly and keep it free of broken leaves and flowers.

Figure 8-1F A floral cooler is used to store flowers. *(Courtesy Steven E. Newman)*

Name: ______________________________ Date: ______________

Exercise 8-2: Constructing a One-Bloom Rose Bud Vase

(by Charles Griner)

Purpose

Step-by-step instructions on how to construct a one-bloom rose bud vase.

Materials

1 longstemmed rose of any color
3 stems of sprengeri or ming foliage
1½ yards of number 3 satin ribbon
2 20- or 22-gauge wires
1 stem of baby's breath (optional)
Leaf polish
Bud vase

Procedure

1. Fill a bud vase with preservative water to within 2 in. of the top.
2. Remove all foliage that will be below the water level. To take off thorns, a rose stripper will be useful (Figure 8-2A).
3. Cover the rose bloom with your hand, and spray the foliage with leaf polish to make it shine.
4. Cut the flower stem below water to a length ½ to 2 times the height of the vase.
5. Wire the rose with a 20- or 22-gauge wire using the straight wire method. Gradually wind the wire around the stem of the rose, being careful not to break off leaves.
6. Insert the flower in the vase and turn the "face," or most desirable side, toward you.
7. Cut three pieces of sprengeri or ming foliage to different lengths and place these so that they arch slightly away from the rose (Figure 8-2B).
8. Add a sprig of baby's breath so that it surrounds the rose inside the foliage.
9. Tie a bow about the size of a corsage bow. Leave three streamers cut to different lengths up to six inches long. Tie the bow using a 20-gauge wire. If smaller wire is used, attach it to a wooden pick.
10. Insert the bow in the front center of the bud vase slightly above the rim. Adjust the streamers.
11. Attach a card to the ribbon or insert a cardett and card (Figure 8-2C). A cardett is a plastic stem with a three-prong holder at the top to hold a card.
12. Refrigerate the completed bud vase.

Name: ______________________________ Date: ______________

Figure 8-2A Use a rose stripper to remove thorns and foliage. *(Courtesy Steven E. Newman)*

Figure 8-2B Add sprengeri or ming fern to the bud vase. *(Courtesy Steven E. Newman)*

Figure 8-2C Attach a bow to the bud vase. *(Courtesy Steven E. Newman)*

Name: ______________________________ Date: ______________

Exercise 8-3: Constructing a Three-Bloom Carnation Bud Vase

(by Charles Griner)

Purpose

Step-by-step instructions on how to construct a three-bloom carnation bud vase.

Materials

3 carnations
5 stems of leatherleaf fern
1 large sprig of baby's breath
Floral preservative
Leaf polish
1½ yards of number 3 satin ribbon
20- or 22-gauge wire
A large-necked bud vase

Procedure

1. Prepare floral preservative, and fill the bud vase to within 2 in. of the top.
2. Spray leatherleaf fern with leaf polish.
3. Wire the carnations using the straight-wire method. Flowers may not need wiring if the stems are strong.
4. Select the smallest carnation and cut it to twice the height of the container.
5. Select the medium-sized bloom and cut it 1 to 2 in. shorter than the tallest flower.
6. Cut the largest flower 1 to 2 in. shorter than the middle flower.
7. Remove all foliage from the stems that will be below the vase line.
8. Insert the flowers in the bud vase. Place the smallest flower in the center, facing up. Stagger the medium flower and place the largest flower at the bottom center (Figure 8-3A). Gradually face the middle and bottom flowers toward the viewer.
9. Select two pieces of leatherleaf which are slightly taller than the flowers, and place these back to back behind the flowers. This causes the leatherleaf to stand up straight behind the flowers.
10. Insert shorter pieces of leatherleaf on each side at a 45° angle and in front of all other stems. An optional piece may be inserted vertically in front of the tallest stem (Figure 8-3B).
11. Insert baby's breath and add a bow centrally, slightly above the front rim of the vase (Figure 8-3C). The bow is attached to a wooden pick that is inserted into the bud vase.
12. Add a card and refrigerate.

Name: ______________________ Date: ______________

Figure 8-3A Placement of the carnations in the bud vase. *(Courtesy Steven E. Newman)*

Figure 8-3B Add leather-leaf to the bud vase. *(Courtesy Steven E. Newman)*

Figure 8-3C Add baby's breath and attach a bow. *(Courtesy Steven E. Newman)*

Name: ______________________________ Date: ______________

Exercise 8-4: Testing Floral Preservatives

(by Charles Griner)

Purpose

To determine how effective floral preservatives are in extending the vase life of flowers.

Materials

Three pint jars
Floral preservative
Six fresh flowers
Knife
Fresh water

Procedure

1. Mix a small amount of water and floral preservative according to the directions on the container.
2. Select two pint jars. Fill one jar half-full with the water/preservative solution and the other half-full with clean tap water.
3. Using a sharp knife, cut six fresh flowers underwater with a knife. Place three of the freshly cut flowers in each of the pint jars.
4. Check the flowers for bruised or damaged spots that might bias the experiment. Replace the flower if any damage is found.
5. Observe the water in the jars for cloudiness and the flowers for signs of deterioration for the next ten days. Record your observations on the observation sheet as instructed.

Observations

1.Record your daily observations of the cut flowers on the following table.

	Water Observations		Flower Observations	
Date	Untreated Water *	Treated Water *	Untreated Water **	Treated Water **

** Rate the condition of the water using a pint jar of fresh water as a check. Rate the cloudiness of the water on a scale of 1 to 10 with fresh water being a 1 and water too cloudy to see through as a 10.*
*** Rate the condition of the flowers on a scale of 1 to 10 also with fresh newly cut flowers being a 1 and a deteriorated, unusable flower as a 10.*

Name: ______________________________ Date: ______________

2. Answer the following questions:

a. How many days lapsed between the start of the experiment until cloudiness was observed in the untreated water? ______________

b. How many days lapsed between the start of the experiment until flower deterioration was observed in the flowers placed in the untreated water? ______________

c. How many days lapsed between the start of the experiment until cloudiness was observed in the treated water? ______________

d. How many days lapsed between the start of the experiment until flower deterioration was observed in the flowers placed in the treated water? ______________

Conclusion

What effect did the floral preservative have on the vase life of the flowers?

SECTION

9

Postharvest Handling and Marketing of Potted Plants and Cut Flowers

Notes

Name: ______________________________ Date: ______________

Exercise 9-1: Effective Selling Skills

(by Charles Griner)

Purpose

To understand the methods and importance of effective selling skills.

Materials

Not applicable to this exercise.

Procedure

Effective selling skills have to be learned. Professional sales people go through a number of steps in helping a customer make a purchase.

Step 1. *The Initial Approach or Greeting* Generally three methods may be used to approach customers in the retail florist: the service, greeting, and merchandise approaches.

The service approach is best used when it is obvious the customer is in a hurry. In this approach, the salesperson might ask the customer if he or she needs assistance by asking, "May I help you?" In many situations this approach is not effective because it usually results in a negative response such as, "I'm just looking." When this happens, you lose control of the sales situation. If customers are in doubt about a purchase, they may leave without giving you a chance to help them with the decision.

In the greeting approach, the salesperson simply welcomes the customer to the store. Use the customer's name if possible with an appropriate greeting, such as, "Good morning." If you know the person, make a personal remark if possible. This approach begins conversation, and you can begin to discover the customer's needs.

In the merchandise approach the salesperson makes a comment about an item that has the customer's attention. To use this approach, the customer must stop to look at a specific item (Figure 9-1A). This approach is usually the most effective because it focuses attention on the merchandise and gives you the opportunity to tell the customer something about it.

Step 2. *Determining Needs* Customer needs are directly related to buying motives. The American Floral Service, Inc. has identified three reasons why customers buy flowers:

1. To fulfill an emergency need.
2. For self-satisfaction and personal enjoyment.
3. To receive personal recognition.

The main reason people buy flowers is for personal recognition. The salesperson's job, therefore, is to discover the level of recognition the customer is seeking. By listening and asking probing questions, this information can be obtained. For example, a man sending flowers to an important client is seeking a high level of recognition. In order to help him make a purchase, you need this information.

Name: ______________________________ Date: ______________

Figure 9-1A One approach to getting a customer is to make a comment about an item that has the customer's attention. *(Courtesy Steven E. Newman)*

Step 3. ***Product Presentation*** After you have determined the reason a customer is buying flowers and the level of recognition the customer is seeking, you should be able to select a few items that might match those needs. Base your suggestions on product, emphasizing the benefits. Include, but don't emphasize, price.

To avoid overwhelming the customer, show no more than three products at a time. If you are not sure about price, show one item from three different price ranges. For example, you might show arrangements for $20.00, $25.00, and $30.00. Explain the added benefits of the higher priced item. People are usually willing to pay more if they are assured of the value of the item. This procedure is called upselling and often results in a larger sale. Remember to emphasize product, not price. If possible, get the customer involved in the sale. Encourage the customer to hold or smell specific flowers, rather than looking at them in a cooler (Figure 9-1B).

Use descriptive language to explain how an arrangement will meet the customer's needs. Avoid such words as "nice and cute," using terms like "fragrant" and "lasting quality," instead.

To present a product well, you must be well informed. This requires effort. Make notes each morning about the flowers on hand and product ideas to use in sales presentations.

Step 4. ***Handling Questions and Objections*** Objections and questions should be viewed as positive feedback. Answering questions gives you the opportunity to present more information. If a customer seems to like one arrangement but has some reservations about making a decision, suggest some changes that could be made to better suit the customer's needs. If the price range of the arrangements you have shown is too high or too low, put those aside and show other arrangements in the correct price range. Show empathy toward the customer and offer product and price alternatives.

Step 5. ***Closing the Sale*** This calls for the customer to make a decision. When you ask for a buying decision, base the decision on product, not price. Ask, for example, if the customer would prefer the arrangement of lilies or the arrangement of mixed spring flowers.

Name: ______________________________ Date: ______________

Figure 9-1B Get the customer involved in the sale by holding or smelling flowers. *(Courtesy Steven E. Newman)*

If the customer is still not ready to make a decision and offers objections to both arrangements, quickly regroup and make suggestions that are more appropriate to his or her needs.

Step 6. *Suggestion Selling* This encourages the customer to buy additional merchandise related to the original purchase. For example, a customer has purchased an arrangement for a birthday party. You might suggest balloons to go along with the arrangement. This is a good way to increase sales.

Step 7. *Writing up the Order* This order should include a brief description of the arrangement. It's a good idea to repeat the description to the customer so that there are no misunderstandings. Confirm the delivery date and the message to be included on the card. Itemize any additional charges such as delivery and sales tax. Give the customer the total charges.

If the customer does not have an established charge account, ask the customer how he would prefer to pay for the order.

Step 8. *Reassuring and Thanking the Customer* This is the last step of a sale. Help the customer feel confident that he or she has made a wise purchase. Thank the customer for selecting your shop and ask the customer to call again.

Name: ______________________________ Date: ______________

Exercise 9-2: Creating a Newspaper Advertisement

(by Charles Griner)

Purpose

To learn how to prepare a newspaper advertisement.

Materials

Old issues of horticulture magazines
Scissors
Paste
Clip art (if available)

Procedure

A good advertisement should have five major elements. These are headline, copy, illustration, price, and business identification. The headline should be something that catches readers' attention and make them want to know more. The copy is the selling message of the advertisement. The copy should make the customer want to own the product or service. The illustration should attract customer attention and create interest. It should also show the customer what the product is like. The price allows the customer to compare the featured item with a similar product. The business signature gives the business name, location, and other information about the business.

Examine the following advertisement. Circle and label each of the five major elements.

Name: ______________________________ Date: ______________

Using information from old issues of horticultural magazines and clip art (if available), create a 5-in., 2-column newspaper ad for a nursery or garden center. The ad should feature a sale on hanging baskets that the nursery is planning. When you have completed the advertisement, check to make certain all five of the major elements have been included.

GLOSSARY

A-REST® a growth retardant used to keep plants short

abscission term used to describe the natural process of leaves falling from the plant

absorption the uptake of nutrients from the soil solution

acclimatization (or acclimation) the process by which plants adapt to another climate or environment

acid organic matter an organic material that leaves an acid residue in the growing medium; an example is sphagnum peat moss

acid reaction fertilizer a fertilizer that causes the soil to become more acid; this type of fertilizer is also acid residue fertilizer

acrylite acrylic a plastic greenhouse covering consisting of two layers of material separated by air space to provide insulation

adult stage of growth advanced period of plant growth, leads to flowering and production of seed

aerated steam steam into which a stream of air is injected to hold the temperature at a constant value, usually 140°F (60°C)

aeration a characteristic condition of soils, good aeration occurs in loose, open soils; poor aeration occurs in heavy, tight soils

aerial roots roots that remain in the air and exchange oxygen and carbon dioxide; they do not absorb water unless they enter a medium or fasten themselves to a solid surface

airfreight transporting products by commercial air carriers; cut flowers are often shipped by airfreight

air pollutant material in the air that is polluting; a few pollutants are sulfur dioxide, ozone, and PAN

alkaline reaction fertilizer a fertilizer that causes the soil to become more alkaline; this type of fertilizer is alkaline residue fertilizer

alloy a combination of several metals. Aluminum alloys are used in the greenhouse framework

anaerobic lives without oxygen

anaphase the period when the chromosomes divide into two equal groups and each half moves to opposite ends of the cell

anion a negatively (–) charged ion; nitrate is an anion NO_3

anion exchange capacity the chemical activity of soils associated with negatively charged ions

annual plant a plant that completes its life cycle in one growing season

anther pollen-bearing structure that is found at the end of the filament

AquaGro® a surfactant (wetting agent) used to reduce the surface tension of water

asexual propagation (vegetative) this method of propagation uses sections of plants that are rooted

aspen fibers shredded wood fibers used to make up an evaporative pad

aspirated thermostat thermostat over which air is blown by a small fan; this type of thermostat is usually more accurate than nonaspirated thermostats

asteraceae one of the largest plant families, including sunflowers, daisies, dandelions, chrysanthemums, and many other plants

autoclave a method of sterilizing objects under pressure

automation mechanical or electronic means of doing work that formerly was done by hand—for example, thermostatic control of a heating system versus manual control

axillary bud a bud that develops in the angle (axil) the leaf makes with the main stem

azalea pot a three-quarter size pot; the height is three-quarters of its width

bent neck a condition of roses that occurs a day or two after cutting; the stem fails to transport water to the flower, and the bud bends over without opening

biennial plant a plant that requires two growing seasons to complete its life cycle

B-I-N paint a paint that is sometimes used successfully to cover pentachlorophenol-treated woods

black cloth a means of providing artificial short days so that short-day plants will bloom

blindness the failure to form flower buds because of a number of factors

B-NINE SP® a growth retardant applied as a foliar spray

bogotá the capital of Colombia, South America, is the center of carnation production in that area

boiler horsepower method of rating the size of a boiler. One boiler horsepower is equal to the heat output of 33,475 Btu per hour

botrytis a disease of the foliage caused by *Botrytis cinerea*

bottom heat heat applied to the rooting or propagation medium to hasten growth

bouyoucos bridge an instrument that measures the moisture content of a growing medium; it is used with a Bouyoucos block

bract a modified leaf that may be highly colored; poinsettias have very colorful red bracts

broadcast sowing a method of sowing seed by sprinkling it randomly over the surface of the medium, as opposed to sowing in rows

brush-cut pruning roses are cut back to a uniform height in one cutting; the pruning is usually done after the Mother's Day cut

Btu (British thermal unit) amount of heat needed to raise the temperature of 1 pound of water 1 degree Fahrenheit, at, or near, its point of maximum density (39.1°F of heat). (Calorie. The amount of heat needed to raise the temperature of 1 gram of water 1°C, at 15°C.)

bud blasting the death of the developing flower bud, usually because of water stress

bud-cut flowers cut in the bud stage and then opened in a special solution

budding a method of grafting used to start roses

bud union the location where the bud was inserted into the understock

bulb cellar room used to give bulb crops a cold temperature treatment; also called a rooting room

bulblet a small daughter bulb of a bulb-producing crop

bushel a volume of media containing 1.25 cubic feet (35.7 l)

calcined clay clay mineral heated to 800°C and then ground into granules

calyx made up of the sepals

calyx splitting the calyx of the carnation breaks open, causing the petals to hang out; "splits" bring low prices in the market

cambial meristem tissues within the stem that divide and cause it to increase in circumference

capillary mat a fiber mat that is used to distribute water to potted plants; the plants take up the water through capillary action

capillary tubes fine water-conducting tubes found naturally in soils

carbon dioxide a colorless, odorless, tasteless gas used by plants to make food sugars (carbohydrates)

carbon monoxide a colorless, odorless, poisonous gas that is formed by burning acrylic plastics in insufficient air

carotenoids yellow, orange, or red pigments found in the plants

cascade petunias petunias that have a trailing habit of growth; these plants are especially suitable for window boxes and hanging baskets

cation a positively (+) charged ion; calcium is a cation (Ca^{++})

cation exchange capacity a chemical property of soils; cations have a positive (+) charge

CEC (cation exchange capacity) chemical activity in which soils pick up and release nutrient elements having a positive charge

cell the smallest basic unit of life

cellulose the wood fraction of bark products

Cfm (cubic feet per minute) a measure of the amount of air moved by a fan; for example, an 18,000-cfm fan moves 18,000 cubic feet of air per minute

chamber steaming this process is the same as vault steaming; items to be steamed are put in a closed chamber

chapin trickle tube a watering system developed by Richard Chapin of Watertown, New York

chelated iron a chemical form of iron that does not become tied up in the soil but remains available for plant uptake

chimney effect movement of air in a greenhouse created when both the top ridge and side ventilators are ope

chlorophyll the green pigment within the cell that is needed to carry on photosynthesis

chloropicrin (tear gas) a chemical used to fumigate soils

chloroplast a plastid containing chlorophyll

chlorosis loss of normal leaf color due to a deficiency of an element or because of root injury

christmas pinch rose plants are pinched to bring the crop into flower for Christmas sales; the soft pinch is made October 17–21; the hard pinch is made October 20–25

chromosomes the structures in the nucleus that contain the genes

clay pot a pot made of fired clay; it is burnt orange in color and fragile (it is easily broken if dropped); it can be steam sterilized, has porous sidewalls, and is heavy in weight

clock auction a wholesale floral auction where the sales are controlled by means of a large clock; clock auctions are located in Aalsmeer and other cities in the Netherlands. In North America, clock auctions are located in Toronto, Montreal, and Vancouver, Canada, and in San Diego, California

clubbed spray a spray-type chrysanthemum having flowers too close together

coated seed seed that is coated by the supplier with clay or other substances; seeds treated in this manner are also known as pelleted seed

cold frame protective structure used to start plants early in the spring; usually covered with glass or plastic

cold storage the treatment given to plants and bulbs to cause certain internal chemical changes that enable them to respond to forcing treatments

community pot/flat a pot/flat where a large number of small seedings are transplanted for a short period before they are moved to other containers

complete fertilizer a fertilizer that contains nitrogen, phosphorus, and potassium, the three major nutrient elements

conduction loss of heat through the covering of the greenhouse and the structural units exposed to the air

connected greenhouse greenhouses that are connected at the side wall

consignment one method used to sell flowers; the grower "consigns" a crop to the wholesaler who sells it and deducts a percentage of the selling price as a commission

constant fertilization the application of a small amount of fertilizer at each watering

container the object in which plants are grown, such as a pot, basket, bucket, or market pack

controlled release fertilizer a fertilizer that releases its nutrients slowly over a predetermined amount of time

convection movement of heated air due to natural forces

cormels small daughter structures produced by a corm; gladiolus produce cormels

cornell cut flower food solution a standard bud-opening solution consisting of 8-HQC and sugar

cornell peat-lite mix a combination of vermiculite or perlite and peat moss; developed at Cornell University and used as a growing medium

corolla all of the petals together

cortex plant tissue just below the epidermis of the root and stem

cotyledon leaves the first seed leaves that develop

creosote a wood preservative that is phytotoxic and should never be used in the greenhouse

critical day length the number of hours at which a plant responds to the day length by undergoing a change in its morphology (changing from vegetative growth to flowering growth)

crown a group of closely grouped stems or plantlets; African violets form multiple crowns

CTF (controlled temperature forcing) a method used to precool lilies

cubic foot a volume of medium with dimensions of one foot by one foot by one foot

cubic yard a volume of medium that is 1 yard long, 1 yard high, and 1 yard wide; 1 cubic yard contains 27 cubic feet and is equal to 21.6 bushels

cultivar a propagated variety of a plant

curtain wall a low sidewall, two to three feet high, made from wood, poured concrete, cinder block, glass, or plastic

curvilinear Greenhouse with curved roof bars, usually at the lower part of the roof

cut flower food Combination of chemicals made up as solution to increase the lasting life of cut flowers

cuticle the protective coating on the leaf

cutin Waterproof, waxy material that makes up the cuticle

Cycocel® a growth retardant that is applied as a soil drench or a foliar spray

cytoplasm the total cell except the nucleus

damping off Death of seedlings due to a soil-borne microor-ganism

day-neutral plant a plant that flowers regardless of the length of day

day temperature the temperature used in the greenhouse from 7 A.M. or 8 A.M. until 5 P.M.

deficiency a level of nutrition that is below that needed for good growth

deionized water Water from which all ions are removed by means of the ion-exchange process

desiccation Excessive drying of flowers; they do not recover from this extreme drying

detached greenhouse a structure that stands apart from another greenhouse

DIF an abbreviation to designate the difference between day and night temperatures. DIF is calculated by subtracting the night temperature from the day temperature. A positive DIF is when the night temperature is less than the day temperature. A negative DIF is when the night temperature is greater than the day temperature

differentiation *See* **Maturation**

diffusion the movement of water into the roots; diffusion may be active or passive

disbudding removing the unwanted axillary bud to allow one bud to flower

disc (disk) florets florets in the center of the chrysanthemum head

disease organism an organism that causes disease; the organisms may be found in the soil, water, or air

diurnal cycle an up-and-down fluctuation of temperature, day and night

division a method of propagation similar to separation

dormancy the resting stage of seed

dormant-eye rose a budded rose where the "eye" is still dormant

dowfume MC2 a very poisonous mixture of methyl-bromide and chloropicrin used to fumigate soils; salvia and carnations should not be planted in bromine treated soils

drip gutter located below the gutter. This is a small channel that carries away the condensation that runs down the roof bar

dry storage at 31°F a method of storing certain flowers without water for extended periods

dwarf strain a low growing cultivar of a plant

eave the point at the top of the greenhouse sidewall to which the roof bars are fastened

eave height the height of the greenhouse sidewall from the floor to the eave

8-HQC 8-hydroxyquinoline citrate is an additive in cut flower foods that helps to reduce stem blockages

8-HQS 8-hydroxyquinoline sulfate is an alternative additive in cut flower foods that performs the same function as 8-HQC

elbow arm mechanical leverage connector for opening and closing sash-type ventilators

embryo a miniature plant contained in the seed

endosperm and cotyledons Food storage tissues in seed

epidermis the bark or skin of the plant

epinasty a growth condition shown by certain plants (tomatoes, marigolds) to excessive amounts of gas in the atmosphere

epiphyte an air-loving plant that grows best in a very porous, highly aerated medium

equinox When the natural day length is equal to the night; March 21 and September 21 are the spring and fall equinoxes, respectively

ESMIGRAN a trace element fertilizer produced by Mallinckrodt Chemical Corporation

essential element a mineral element that is necessary for the growth of an organism; sixteen elements are considered necessary for the optimum growth of plants

Ethrel® a growth-regulating chemical that is used to stimulate flowering in certain crops

ethylene a gas formed from decaying plants, apples, and also produced by all flowers; it usually reduces the lasting life of cut flowers

ethylene scrubber a material that absorbs ethylene gas from the atmosphere; this material is usually packed in shipping boxes when ethylene is a problem

etiolated seedlings Stretched, leggy seedlings resulting from exposure to a very low light intensity

exchangeable form Describes the status of mineral elements in soils; elements may be exchangeable or non-exchangeable

exhaust fan thirty-six to 48-inch (1 to 1.2 m) diameter fan used to exhaust air from the greenhouse

family a subdivision of the plant kingdom; a family is subdivided into genera

fan and pad cooling a method of reducing temperatures in the greenhouse. Outside air is pulled into a greenhouse through a wet pad by means of a fan

fertilizer a material that contains nutrient elements

fiberglass a plastic that contains fibers of glass that give it strength. Fiberglass may be flat or corrugated. It is used as a greenhouse glazing material. It burns very quickly

fiber pot/pack Pots/packs made of compressed fiber

filament Stalk-like unit holding the anthers

finned pipe Heat pipe to which closely spaced metal fins are fixed to increase the radiating surface of the pipe. One foot (0.3 m) of finned two-inch (5 cm) pipe has the same radiating capacity as five feet (1.5 m) of nonfinned two-inch (5 cm) pipe

fir bark the medium used to grow orchids; it is usually mixed with peat moss

fire-tube boiler a boiler in which the heat is conducted through tubes surrounded by the water to be heated

first true leaves the first true leaves that develop after the cotyledon leaves

five leaflets a rose leaf that has five leaflets

flaccid Only partially filled with fluid, as in a wilted seedling

floral bud development the next stage in the growth of the flower after bud initiation

floral bud initiation the process of changing from a vegetative growing condition to a flowering stage of growth

floral preservative a special solution used to increase the lasting life of cut flowers and to open bud-cut chrysanthemums and carnations. Also called cut flower food

floret an individual flower on a spike, such as a gladiolus floret

flower sheath the part of the orchid from which the flower stalk develops

fluorine an element that exists as a gas or in combination with other elements; in excess amounts it causes plant injury

foam material Retail florists use this material as a base for cut flower arrangements

FOGGIT® nozzle a special nozzle that produces a very fine, foglike spray; it is used for hand watering cuttings or seedlings

foliar analysis (tissue testing) a method of determining the nutrient content of plant tissue by means of a chemical analysis of the leaves or other parts of the plant

footcandle a measure of light intensity. Illumination at a point on a surface which is one foot from and perpendicular to a uniform point source of one candle

forced into bloom the procedure whereby plants are made to bloom according to a schedule; for example, azaleas are forced for Easter sales

formalin commercial formaldehyde used to treat seed flats

foundation the underground structure that supports sidewalls—also includes the footing that supports the load-bearing posts. The footing must be installed below the frost line

freeze-drying a method of preserving cut flowers, fruits, and vegetables in a permanent condition

frit controlled release fertilizers, usually made of soft glass infused with nutrient elements and then ground to a fine powder

frond a branch of a fern or palm plant

FRP fiberglass reinforced polyester

FTE 555 fritted trace element fertilizer

fumigation Injection of chemicals into the soil or other medium to kill diseases and insects

fungicidal drench a chemical drench used to combat certain soilborne diseases

fungicide a chemical used to combat fungus infections

genes the messengers in the chromosomes that carry hereditary information

germination development of a seed from a resting stage to a stage of growth

germination media the media (soil or soilless mixtures) in which seed is germinated

gibberellic acid a growth-stimulating chemical

glazing sill sash sill mounted at the top of the curtain wall, the unit on which the side glass rests. If a side ventilator is used, it is called a sash sill

"gooseneck" stage the stage of growth at which the daffodil flower assumes a position horizontal to the ground

gothic arch greenhouse structure shaped like a church arch

gradual pruning a method used to cut back roses that does not take them out of production completely

grafting a method of propagation that is used for difficult-to-root plants; it is also used to take advantage of special characteristics of the root stock

gravity flow system gravity causes the return flow of cold water to the boiler

greenhouse atmosphere the air inside the greenhouse

"greenhouse effect" process by which the rays of the sun entering the greenhouse as light are changed to heat energy and trapped inside

growing medium a mixture of several materials in which plants are grown

guard cell specialized cells that control the opening and closing of the stomata; found mainly on the undersides of the leaves

guttation the minute loss of water during the night at the tips and edges of the leaves

gutter section between two attached (ridge and furrow) greenhouses. It carries off rain water and melted snow

hanging basket a growing container with attached supports that can be hung from a hook or rack

hardening off a toning and strengthening process in which the temperature is lowered and/or water is withheld from the plants

hard pinch a pinch that removes enough terminal growth that it can be used as a cutting

hardwood barks barks obtained from deciduous trees such as maple and oak

hardwood mulch a hardwood bark product that contains less than 85 percent lignin

heat exchanger part of a heating system that is similar in action to an automobile radiator. Steam or hot water is circulated through a fin coil, and a fan blows air over the coil to force heat into the surrounding area. Unit heaters contain heat exchangers

heat transfer coefficient a numerical value given to the relative amount of heat transmitted or transferred through a material. The heat transfer coefficient of glass is 1.15

herbaceous perennial a plant that lives from year to year; the top growth dies back to the roots after a frost, but the roots live from year to year; a peony is an herbaceous perennial

herbicide weed killers; only certain types of chemicals are safe to use inside the greenhouse; 2, 4-D types should never be stored or used in a greenhouse

HID (high intensity discharge lamps) used to supplement the daylight

homogenous population plants propagated vegetatively are all the same

hotbed a heated cold frame

HOZON® a simple fertilizer proportioner

humidistat a control for the operation of a fogging or misting unit that operates on the relative humidity content of the air

humus organic matter in an advanced state of decay

hybridization plant breeders combine two hybrids to produce an improved cultivar

hydathode the water of guttation exudes from this organ at the leaf tip

hydrophyte a water-loving plant

hydroponic solution nutrient solution in which plants are grown

hypnum peat a peat moss derived from *Hypnum moss*

hypocotyl the first true stem of the seed

incandescent lamps standard bulb-type lights that are high in red and infrared energy

infiltration movement of outside air into the greenhouse through laps in the glass, holes, and other small openings

inflorescence the flowering head of a plant

infrared radiation long-wavelength radiation from the sun, felt as heat

inhibitor a substance in the seed coat that prevents or slows down the germination process

inorganic material material that does not contain carbon; sand, perlite, and vermiculite are inorganic materials

insecticide a chemical used to control insects

intercalary zone In monocotyledonous plants, an area of meristematic tissue located just above the node

intercellular wall the wall laid down in the cell

interior foliage plants tropical foliage plants that are grown in a special environment to condition them to life indoors

intermittent mist a low-pressure system that applies a fine, fog-type mist over cuttings; the frequency of the cycle of misting is controlled by mechanical or electronic means

interrupted lighting a schedule of long and short days used to develop better flower heads on 'Indianapolis' and 'Mefo' cultivars in winter

interveinal chlorosis yellowing of the tissues between the veins

IPM (Integrated Pest Management) the use of all possible means to control greenhouse pests. This includes diseases, insects, and weeds. Excluding insects by screening openings, ventilating to reduce relative humidity to control certain diseases, mechanically removing weeds by hand pulling or hoeing, using biological methods to control insects and diseases are all part of the IPM approach. Chemical pesticides are used only as a last resort

irrigation the act of putting water on a crop

Jiffy-Mix® commercial formulation of the Cornell peatlite mixes made and sold by Jiffy Products of America

Jiffy-7® a commercially prepared, compressed peat moss disk used to propagate plants

juvenile stage of growth early stage of plant growth

Kool-cel® tradename for a commercial, paper-based evaporative pad

lath house a structure covered with wood or metal slats. It is used to protect plants from a high light intensity

layering a method of propagation that produces a new plant while the stem being rooted is still attached to the mother plant; the mother plant serves a life support function. there are several methods of layering with air layering and mound layering popular choices

leaching action the dilution and flushing away of nutrients or high soluble salt levels by means of heavy irrigations

lead newly developing shoot of an orchid *(Cattleya)*

leaf-bud cutting a small section of the stem with a leaf attached; the new shoot arises from the axillary bud

leaf cutting a single leaf from which new plants arise

leaf mold the rotted remains of leaves

leaf scorch a burning or necrosis of leaves; this condition is common to 'Croft' cultivar lilies

leaf tip burn a necrosis or "scorching" of tips and edges of leaves; in members of the lily family, leaf tip burn is thought to be caused by fluorine toxicity; other causes include dry air, high soluble salt levels, and pesticide injury

lean-to greenhouse with a one-sided roof, usually built against another building or a greenhouse

light compensation point the lowest light intensity at which the amount of food produced by photosynthesis just equals the energy used up in respiration

light meter a meter used to measure light intensity; the measurement is usually in footcandles

light saturation point the intensity of light beyond which the plant cannot use a brighter light for photosynthesis

lignin the main component of bark, which decomposes slowly

lily pot a three-quarter size pot

limiting growth factor a factor that limits growth in the greenhouse; for example, a low light intensity in the winter is the most important limiting factor

liquid fertilizer a fertilizer applied as a liquid

lobed deep indentation in the margins of a leaf; for example, a five-lobed leaf

long-day plant a plant that flowers only when the day length is longer than a critical number of hours

louver a mechanically or electrically operated unit that opens and closes to allow air to enter or exit a greenhouse

lumen a measure of the light flux falling on a surface one square foot in area, every part of which is one foot from a point source of light of one footcandle in all directions

luxury consumption the uptake of nutrient elements in amounts greater than is required for growth

macro element minerals needed in large quantities include carbon, hydrogen, oxygen, nitrogen, phosphorus, potassium, calcium, magnesium, and sulfur

Magamp® controlled release magnesium ammonium phosphate fertilizer

makeup air air needed to cause the burner of a heater to operate at 100 percent of its efficiency, usually drawn from outside the greenhouse

manure animal waste that is mixed with bedding material such as straw, sawdust, peat moss, wood chips, or similar products

marginal necrosis death of the margin (edge) of the leaf tissue; *see* **Leaf tip burn**

market pack a fiber or plastic container used to grow bedding plants; the container holds from six to twelve plants

mass market outlet a supermarket chain or discount stores that sell flowers

maturation area of the root where the cells reach maturity and form tissues

maximum the highest level or the greatest value, such as the maximum temperature for growth

mechanical support support given by the growing medium to the plant

medium the term used to describe the "soil" in which plants are grown; the plural of medium is media

meristematic actively dividing tissues

meristem culture a form of tissue culture

meristem tissue culture a method of micropropagating plants using single cells; special techniques are required to grow plants from the cells; the method is also known as cloning

mesophyll the middle tissue layers of the leaf, below the palisade layers

metabolic processes growth processes, or processes that carry on metabolism

metaphase the second phase of cell division in which the divided chromosomes line up at the center, or equator, of the cell

mho a measure of electrical conductance; the term is used to report the level of soluble salt concentration (mho = siemens)

micro element minerals needed in extremely small or trace amounts are iron, manganese, copper, boron, zinc, molybdenum, and chlorine

microorganism a microscopic living organism in the soil or medium

micropropagation (tissue culture) a new propagation technique that uses a single cell of the meristematic tissue of a plant to produce a new plant; antiseptic techniques and specialized growth media are required; the procedure is also known as cloning

milliliter a unit of measurement for liquids; one milliliter (1 ml) = one cubic centimeter (1 cm^3) = one-thirtieth ($^1/_{30}$ oz)

millimho a measure of conductance, one-thousandth of a mho; a term used in reporting total dissolved solids or total soluble salts in solutions (millimho = millisiemens)

millimicron a measure of length; one millimicron equals one twenty-five millionth of an inch

mineral element a substance used by an organism as a source of energy

minimum lowest or least, such as the minimum temperature for growth

mist fertilizer a low-nutrient solution applied to newly propagated vegetative cuttings

mitochondria units within the cell that carry on respiration

mitosis cell division that occurs in several phases

mobility the ability of mineral elements to move from one part of the plant to another; usually the elements move from the older tissue to newly developing tissues

moisture stress a physiological condition caused by a shortage of water

monocotyledon a single cotyledon plant; corn is a monocotyledon

monopodial growth an upright type of growth shown by some orchids; *Vanda* orchids have this habit of growth

motorized accelerated system heating system that uses a pump or pumps to return cold water to the boiler

mottle leaf a speckling and crinkling of the leaf tissue due to a lack of zinc

multicell pak a plastic tray with many cells used to grow bedding plants as individual plants

mutation a change in the character of an organism; mutations may be spontaneous, as in nature, or they may be induced artificially by chemicals or radiation

nanometer a measure of wavelength that is equal to a millimicron

naphthanate, copper and zinc wood preservatives (Cuprinol is one trade name) that are usually safe to use to treat greenhouse wood

negative geotropism a reaction in which a plant bends upward away from gravity

nematode an eel worm that lives in the soil and attacks the roots of plants

neutral reaction fertilizer a fertilizer that has no effect on the soil reaction

night temperature the temperature used in the greenhouse from 5:00 P.M. until 7:00 A.M. or 8:00 A.M.

nitrobacter bacteria bacteria that change the nitrite form of nitrogen to the nitrate form

nitrosomonas bacteria bacteria that change the ammonium form of nitrogen to the nitrite form

nucleus large central part of the cell that is surrounded by a membrane; it contains the chromosomes

nutrient agar a medium used in tissue culture and seed sowing; agar is derived from seaweed (kelp) by a special process, and nutrients are added to it

nutrient mist a mist solution containing a small amount of fertilizer

one-half strength one-half the recommended amount

optimum best level or rate, such as the optimum rate of fertilization. Organic fertilizers are also often considered to be naturally occurring

organic fertilizer a fertilizer material that contains carbon. Organic fertilizers are also often considered to be naturally occurring

organic material material containing carbon; peat moss, sawdust, and wood chips are organic materials

osmocote® controlled release plastic-coated fertilizer

outright purchase a wholesaler buys flowers outright from the grower and sells them to retailers

ovary the flower part that contains the ovule, or the seed

overfertilization the application of too much fertilizer

overpotting putting too small a plant into too large a container

overwatering the application of too much water

oxalic acid a compound found in plants that can cause serious internal injury

palisade layers tissues just below the epidermis of the leaf; most photosynthesis takes place in the palisade layers

palmately compound a leaf form consisting of several leaflets that are attached at the same point on the petiole

panning potting poinsettia plants or bulbs

paradichlorobenzene a moth repellant that should not be used on tulips or other bulb crops to repel mice

parenchyma cells tissue that is made up of loosely packed cells

pasteurization process that kills only harmful disease organisms; aerated steam at 140°F–160°F (60°C–72°C) is used to pasteurize soils

peat bog the area from which peat is dug; a natural deposit of peat moss

peat moss organic material used to grow plants

peat pots pots made of compressed peat moss and paper

"pencil" stage at this stage of daffodil flower stem development, the bud points upright

pentachlorophenol phytotoxic wood preservative that should never be used in the greenhouse

perched water table an area that contains 100 percent water at the bottom of a pot or bench filled with growing medium

perennial plant a plant that lives from year to year; *see* **Herbaceous perennial**

perforated polyethylene tubing plastic tubing eighteen to twenty-four inches (45–60 cm) wide, with regularly spaced holes on each side. Used to distribute air in a greenhouse

PERK® a trace element fertilizer manufactured by Kerr McGee Fertilizer Company

perlite a volcanic mineral expanded by heating to 1,800°F (982°C) used as a soil additive, it is white in color

permanent wilting point the point beyond which a wilting plant will not recover even when it is given water

petals the showy, colored parts of the flower

petiole the lower part of the leaf where it is attached to the plant; the petiole is often a stalk

PH the term used to denote the acidity or alkalinity of media or solutions

phloem the vascular tissue that carries organic foods from the leaves to the roots and from the roots to the leaves

photoperiod the length of the daily light period

photoperiod responder a plant whose flowering depends upon the day length

photosynthates foods produced by the plant as a result of photosynthesis

photosynthesis the conversion of light energy into chemical energy by means of chlorophyll; process by which plants change the energy of sunlight into usable food (carbohydrates and proteins)

phototropism response shown by plants to light coming from one direction

phytochrome an enzyme that controls the response of plants to day length

phytotoxic causing plant damage; certain gases are phytotoxic

pinch and one-half a carnation crop is pinched one month after planting; thirty to fifty days after the first pinch, one-half of the shoots that develop from the first pinch are pinched. The process spreads out flowering over a longer period

pinching the procedure by which a small part of the growing tip of the plant is removed to cause the development of axillary buds

pinnately compound a leaf form consisting of several leaflets attached to a long petiole; the arrangement of leaflets may be opposite or alternate on the petiole

pistil female reproductive part of the plant consisting of the stigma, style, and ovary

pitch the angle the roof makes with the horizontal

plant nutrient material used by plants to make growth

plastic pot a pot made of any one of several types of plastics; it is light in weight, has nonporous sidewalls, and is available in many sizes and colors; plastic pots usually cannot be steam sterilized

plastids small bodies in the cell that contain the chloroplasts

plumule the first terminal bud of the embryo; the first shoot that emerges from the seed

polyethylene a petroleum-based flexible plastic used for many purposes; greenhouses can be covered with polyethylene

post-emergent damping-off Soil organisms kill seedlings shortly after they emerge from the medium

postharvest life the lasting life of a flower after it is harvested from the plant

ppm (parts per million) a measure of the concentration of a fertilizer solution; for example, 200 ppm nitrogen. One ppm equals one milligram per liter

precooling a cold temperature treatment given to tulips and other bulbs to make them flower

pre-emergent damping-off Soil organisms kill seedlings shortly before they emerge from the medium

preharvest handling conditions under which the crop is grown before harvest

premature sprouting a condition that affects lilies, causing them to sprout in the shipping case

prepared hyacinths hyacinth bulbs from plants that were grown in heated soils

preservative, wood a fluid used to treat wood to prevent rotting. Creosote and pentachlorophenol are wood preservatives, but they must not be used in a greenhouse—their fumes are poisonous to plants

primary wall the first wall laid down in the cell

Pro-Mix® commercial formulation of the Cornell peatlite mix made and sold by Premier Horticulture

propagate to reproduce by seed or other means

propagation the increase of a plant species through generation after generation

propagation bench the bench containing the rooting medium into which the cuttings are stuck to develop roots (rooting)

propagation environment the area where the propagation bench is located

propane gas a fuel used to heat greenhouses; it is also burned to add carbon dioxide to the greenhouse atmosphere

prophase the first phase of cell division

proportioner a mechanical unit that dilutes a small amount of concentrated fertilizer stock solution into a large volume of water

protective environment surroundings for growing plants that are protected from the weather. The environment may be provided in a glasshouse, plastic-covered greenhouse, cold frame, or hot bed

protoplasm the organic contents of the cell

pruning removal of extra side shoots

pseudobulbs the specialized storage tissues of orchids that are above the surface of the medium

psi pounds per square inch, a method of expressing steam or water pressure

puddled soil a clay soil that is worked when it is too wet; its structure is ruined

purlin a structural unit that is used to support the sash bars. It is built lengthwise along the greenhouse. Purlins are supported by purlin posts and purlin braces

PVC polyvinyl chloride, a sheet plastic film

PVF polyvinyl fluoride, a fluorinated plastic that resists weathering. "Tedlar" is a commercial product used to coat fiberglass to make it last longer outdoors. PFV is resistant to ultraviolet radiation

pythium a root rot disease of ornamentals; a pre-emergent damping-off disease organism

quonset a type of curved greenhouse that takes its name from the Quonset hut of World War II

rack and pinion gear-type device for opening and closing sash-type ventilators

radiant energy energy that comes from a radiant source such as the sun

radiational cooling the loss of heat from plant leaves by means of radiation through the greenhouse glass; the loss is most intense on clear, cloudless nights

radiator a type of heat exchanger

radicle the first true root that emerges from the seed

ralstonia bacterial organisms that cause bacterial diseases in several plants; there are different species of *Ralstonia*

RA-PID-GRO® a soluble fertilizer with an analysis of 23-19-17; in vegetative propagation it is used at the rate of 4 ounces (113 g) in 100 gallons (378.5 l) of water and is applied as a nutrient mist

ratoon the offshoot pineapple that is used to propagate new plants

ray floret flowers around the perimeter of the chrysanthemum head

receptacle the swollen part of the stem or stalk; the flower parts are attached to the receptacle

Redi-Earth® commercial formulation of the Cornell peat-like mix made and sold by The Scotts Company, Marysville, OH

reed-sedge peat peat derived from the various reeds and sedges that usually grow together

relative humidity the amount of moisture actually contained in the air compared to the total amount that can be held at a specific temperature with no condensation

repotting or shifting up replanting plants into another, usually larger, container

respiration the process of exchanging oxygen and carbon dioxide through the stomata

response group when applied to chrysanthemums, refers to the number of weeks after the start of short days that are required before the plant will bloom; for example, a chrysanthemum may belong to a ten-week response group

retail outlet a place where flowers are sold to the public; it may be a traditional florist or a mass market outlet

rhizoctonia a soilborne disease that causes damping-off of seedlings; the injury occurs at the soil line and is usually post-emergent

rhizome an underground specialized stem

ridge the top of the greenhouse; a ridge cap can be used as a walkway on the outside of the roof. Sash bars and the ridge ventilator fasten to the ridge

ring spot a discoloration of African violet leaves resulting from cold water spotting the leaves; the cold water kills the chlorophyll

roof bar or sash bar a specially grooved bar of wood or aluminum, which is used to hold the greenhouse covering

roof bar cap a special unit that is fastened to the top of the roof bar to prevent glass slippage and leaks and to reduce weathering of the glazing compound

root cap a group of cells at the tip of the root that protect the root

rooted cutting the vegetatively propagated part of the plant that develops roots

root hairs specialized root cells in the epidermis of the root that absorb water and nutrients

rooting hormone a root-stimulating substance that is applied to the base of a cutting; rooting hormones are available in several strengths

rosa manetti the understock commonly used for grafting roses

salt index a measure of the amount of salts a fertilizer adds to the soil

saran a polypropylene mesh material used to cover structures to reduce the light intensity. Mostly used in Florida and other high light areas

sash bar *See* **Roof bar**

scarification the process of scratching a hard seed coat so that it can take up water

scion the upper part of a grafted plant; the scion is fastened to the root stock

secondary roots the branched roots that develop from the initial roots formed

seed geraniums geraniums grown from seed

seedling newly developing plant, started from a seed

semi-iron a greenhouse framework that combines metal and wood as structural units

senescence the change of cut flowers from maturity to death

sepals outermost parts of the flower

separation a method of propagation in which clumps of roots or bulbs are separated

sexual propagation this procedure uses seeds of plants for producing new plants

shading the treatment given to mums to cause them to flower when the natural days are longer than the critical day length

shearing a method of cutting back, or heading, plants and shaping them; azaleas are sheared

sheet film plastic flexible plastic that comes in very thin sheets; also known as flexible plastic. Polyethylene is a sheet film plastic

short-day plant a plant that flowers only when the day length is shorter than a critical number of hours

sideposts the weight-bearing units of the greenhouse

simazine a weed killer considered to be safe for use in mature terrestrial orchids; it is not safe to use on seedlings

single-pinch carnations carnations that are given one pinch

single stem a crop grown for single stems produces only one flower to a stem

sleepiness the premature closing of carnation flowers, usually due to ethylene gas exposure

social necessity a social event that is considered necessary to life, such as weddings, births, deaths, and illness

softened water water in which the calcium and magnesium ions and carbonates are exchanged chemically for sodium ions

soft pinch the method of removing a small part (1/2-inch [1.25 cm]) of the top of the plant

softwood barks barks obtained from conifers such as fir, pine, and redwood

softwood mulch a softwood bark product that contains less than 85 percent lignin

soil solution the liquid part of the growing medium that contains the dissolved nutrients

soil structure refers to the arrangement of the soil particles

soil testing a method of determining the nutrient content of a growing medium

soil texture refers to the size of the soil particles; gravel, sand, silt, and clay are classes of texture

solar energy radiant energy from the sun

soluble salts the residue of fertilizers in the soil; soluble salts may be left by other materials, such as manure

Solubridge® an instrument that measures the total soluble salt content of growing media or solutions

sorus a fruiting body on the fronds of ferns that emits spores; sori is the plural of sorus

Speedling® geraniums seed geraniums grown by the Speedling Corp. of Florida

sphagnum moss peat a particular type of peat moss derived from the Sphagnum moss species

spike a type of floral inflorescence

"spitting" the breaking out of the flower stalk of the hyacinth

spray-type mum chrysanthemums grown with many stems attached to the single large stem; also known as pompons

stage G the stage in the development of tulip bulbs when all of the floral parts are visible

stamen the male reproductive part of the plant, consisting of the filament and stamen

standard grades designations used to sort cut flowers to standard stem lengths or other measures of quality

standard mum a large-flowered chrysanthemum, the "football" mum

standard pot a full size pot for which the height is equal to the width, measured at the top of the pot

started eye rose a budded rose plant with an "eye" growing as a branch

steam sterilization the use of steam to sterilize the soil, pots, tools, and other materials used to produce a crop. A temperature of 160°F (57°C) is held for thirty minutes

stem section cutting a section of the stem, usually with a leaf attached

sterilization the process used to kill all organisms in a medium; may be accomplished by steam or chemicals

stigma the part of the pistil that receives the pollen grain

stipules leaf-like growths at the base of petioles; geraniums and roses have stipules

stock plants a group of plants maintained to produce vegetative propagation material (cuttings)

stoma an opening in the leaf or epidermis that is controlled by guard cells; the stomata (plural of stoma) are the sites of gaseous exchange between the inside of the leaf and the atmosphere

style the stalk through which the pollen tube grows to fertilize the ovary

styrofoam a type of plastic used to make lightweight pots

subirrigation flats and/or pots are watered from the bottom, and the water rises by capillary action

sublimed the change of substance from a solid state to a gaseous state without passing through an intermediate state; carbon dioxide sublimes from dry ice to the gas

surfactant a chemical used to reduce the surface tension of water

Surfside® surfactant (wetting agent) used to reduce the surface tension of water

sympodial growth the horizontal growth of a rhizome; *Cattleya* orchids have this habit of growth

telophase the formation of new daughter cells

tendrils long, twining appendages used by plants for support

tensiometer a mechanical device that measures the moisture content of a growing medium

terminal cutting the most commonly used method of propagation, in which the stem tip is rooted

termite a wood-eating insect that may live underground or inside wood; it uses wood for food and destroys it

terrestrial a plant that grows in soil as compared to an epiphytic plant, which grows in air

testa the outer part of the seed coat

thermal blanket a movable cover that is pulled at night to reduce the heat loss from the greenhouse

thermostat a device that measures the temperature and controls the operation of a heating unit

thielavlopsis basicola an organism that causes root rot in pansies, poinsettias, and geraniums

thomas method of steaming surface steaming methods developed by Mr. John Thomas, snapdragon grower of Whitford, Pennsylvania

tip burn *See* **Leaf scorch**

topsoil Soil high in organic matter, usually the top six-inch (15 cm) layer of soil

totipotency the ability of each cell in a plant to reproduce the entire plant as an exact copy of the parent; in tissue-culture techniques, this ability is known as "cloning."

traditional retail florist a florist whose major business consists of the sale of cut flowers, arrangements, pot plants, and some accessories

translocation the movement of water and organic foods through the vascular tissues

translucent a term used to describe the light-transmitting property of a material. Translucent white plastic allows light to shine through, but the light is diffused and images cannot be seen clearly through the plastic

transmittance (light) the passage of light through a greenhouse covering. Black plastic has a light transmittance of 0 percent. Glass may transmit 92 percent of the light that strikes it, fiberglass about 90 percent

transpiration the loss of water in the form of vapor from the leaves; transpiration takes place through the stomata

transplanting moving plants from one container to another; usually the container size increases; also, shifting seedlings from seed flats to packs

trimming for steam procedure whereby the water level in a hot water boiler is lowered slightly so that steam can be produced for steam sterilization

trombone heat system a continuous line of heat pipes that look like a trombone

tropical foliage plants plants originating in tropical regions that are grown for their foliage

truss a member used to support the greenhouse

tuber a specialized stem structure that serves primarily as a food storage organ; white potatoes are tubers

tuberous root fleshy root that acts as a storage tissue

tulip topple a physiological disease of tulips in which a lack of calcium causes the stem to collapse

turgid filled or swollen with fluid

ultraviolet inhibitor a chemical added to polyethylene film that causes the plastic to resist breakdown due to the ultraviolet rays of the sun

ultraviolet radiation short-wavelength radiation of the sun; ultraviolet radiation causes sunburn in humans

understock (rootstock) the lower part of the grafted plant

uneven span greenhouse the roof bars on one side are longer than those on the other; as a result, the roof has an uneven span

unicellular made of one cell

unit steam heater a small heating unit usually mounted overhead in a greenhouse

vacuoles a large empty chamber in a cell that is surrounded by a membrane

vapam a soil fumigant

variegated shades of dark and light colors

vascular tissue the specialized tissues that make up the pipelines that conduct water and nutrients through the plant

vegetative nonflowering stage of growth

vegetative meristem nonflowering growing point of the plant

velamen the covering of aerial orchid roots

ventilator a movable unit fastened to the ridge and side-wall, used to ventilate the greenhouse on hot days or whenever temperatures go above recommended levels

ventilator header the ventilator rests on the ventilator header

vermiculite a mica mineral expanded by heating to 1,400°F (760°C) used as part of the peat-lite mixes

verticillium a fungus-producing organism that grows in the vascular system of the plant and causes wilting

verticillium wilt soilborne disease

virus extremely small organisms that cause a variety of plant diseases

volatiles fumes given off from paints or other materials; some are damaging to plants

waterlogged soil soil that is overwatered

water tube boiler a boiler in which the water to be heated is conducted through tubes that are surrounded by the heat rising from the fire

watt the unit of measurement of electrical energy consumed

wet-cold storage storage of cut flowers with the stems inserted in a holding or cut flower food solution; the storage temperature is 33°F–35°F (0.5°C–1.5°C)

wetting agent material added to water to reduce the surface tension and make the water wetter

WF&FSA wholesale florists and Florist Suppliers of America

wholesale distributor buys flowers from the grower and sells them to the retailer

witches' broom the unchecked growth of many shoots caused by a lack of boron in chrysanthemums, carnations, and roses; the resulting growth is broom-like in appearance

xylem the vascular tissue that carries water from the roots to the leaves

year-round flowering the practice of flowering certain plants year-round; for example, azaleas and chrysanthemums are flowered year-round

zoned heating a large greenhouse is equipped with several thermostats to control the temperatures by "zones." A more uniform temperature is possible with zoned heating than if one thermostat is used

CPSIA information can be obtained
at www.ICGtesting.com
Printed in the USA
BVHW012343250620
582113BV00002B/1

9 781418 030834